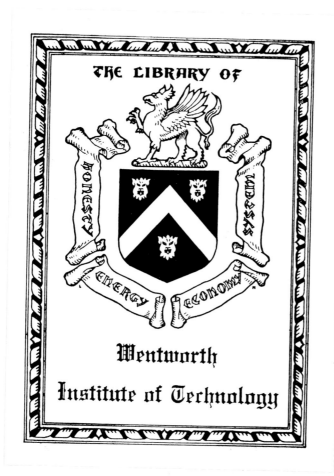

Programmable Controllers
for Factory Automation

MANUFACTURING ENGINEERING AND MATERIALS PROCESSING

A Series of Reference Books and Textbooks

SERIES EDITORS

Geoffrey Boothroyd

Chairman, Department of Industrial and Manufacturing Engineering
University of Rhode Island
Kingston, Rhode Island

George E. Dieter

Dean, College of Engineering
University of Maryland
College Park, Maryland

1. Computers in Manufacturing, *U. Rembold, M. Seth and J. S. Weinstein*
2. Cold Rolling of Steel, *William L. Roberts*
3. Strengthening of Ceramics: Treatments, Tests and Design Applications, *Henry P. Kirchner*
4. Metal Forming: The Application of Limit Analysis, *Betzalel Avitzur*
5. Improving Productivity by Classification, Coding, and Data Base Standardization: The Key to Maximizing CAD/CAM and Group Technology, *William F. Hyde*
6. Automatic Assembly, *Geoffrey Boothroyd, Corrado Poli, and Laurence E. Murch*
7. Manufacturing Engineering Processes, *Leo Alting*
8. Modern Ceramic Engineering: Properties, Processing, and Use in Design, *David W. Richerson*
9. Interface Technology for Computer-Controlled Manufacturing Processes, *Ulrich Rembold, Karl Armbruster, and Wolfgang Ülzmann*
10. Hot Rolling of Steel, *William L. Roberts*
11. Adhesives in Manufacturing, *edited by Gerald L. Schneberger*
12. Understanding the Manufacturing Process: Key to Successful CAD/CAM Implementation, *Joseph Harrington, Jr.*
13. Industrial Materials Science and Engineering, *edited by Lawrence E. Murr*
14. Lubricants and Lubrication in Metalworking Operations, *Elliot S. Nachtman and Serope Kalpakjian*
15. Manufacturing Engineering: An Introduction to the Basic Functions, *John P. Tanner*
16. Computer-Integrated Manufacturing Technology and Systems, *Ulrich Rembold, Christian Blume, and Ruediger Dillmann*
17. Connections in Electronic Assemblies, *Anthony J. Bilotta*

OTHER VOLUMES IN PREPARATION

Programmable Controllers for Factory Automation

David G. Johnson

Automation Controls Department
General Electric Company
Charlottesville, Virginia

Marcel Dekker, Inc.   New York and Basel

Library of Congress Cataloging-in-Publication Data

Johnson, David G.,
    Programmable controllers for factory automation.
    (Manufacturing engineering and materials processing ;
20)
    Bibliography: p.
    Includes index.
    1. Production engineering. 2. Production management.
3. Programmable controllers. 4. Automation. I. Title.
II. Series.
TS176.J63 1987   670.42'7   86-29053
ISBN 0-8247-7674-7

MARCEL DEKKER, INC.
270 Madison Avenue, New York, New York 10016

Current printing (last digit):
10 9 8 7 6 5 4 3 2 1

PRINTED IN THE UNITED STATES OF AMERICA

Preface

Like all good ideas, the idea behind the development of programmable controllers came from the need to do something better.
That something was, in the late 1960s, the control of automobile
manufacturing processes at the Hydra-matic Division of General
Motors in Detroit, Michigan. The traditional control schemes involved the use of hundreds, sometimes thousands of electromechanical relays, wired together with thousands of feet of wire
in a very complex manner. These control "panels" as they were
called, worked reasonably well until one or more of the relays involved failed, or a change in the manufacturing process dictated
the reworking of the control panel. A failed relay was often difficult to diagnose, and reworking the existing panel to adapt to a
different process was normally so expensive that the automotive
companies usually discarded the old panel in favor of a newly designed one. The development of the programmable controller
addressed both of these problems through improved reliability and
diagnostics capability, and the flexibility required to change the
control "logic" simply by changing the software in the system
rather than reworking or replacing the control panel.

After the early acceptance in Detroit, the use of programmable controllers grew rapidly and spread to many other discrete
part manufacturing industries and later to batch and continuous
process related industries. With the introduction of the micro-

processor, programmable controllers began to collect and compute small amounts of data related to the manufacturing process they were controlling. Later, sophisticated methods were developed for the user of a programmable controller to interact with the system. These included touch-sensitive cathode ray (CRT) and voice interaction. With the growing acceptance of computer-integrated manufacturing (CIM), programmable controllers are now being arranged in communication networks with other intelligent controllers and computer systems to form manufacturing cells and flexible manufacturing systems.

During my experience with General Electric, I noticed that while there were many training-related documents on programmable controllers, most of these were designed to be used with a specific manufacturer's equipment and were fairly narrow, relating to a specific functional area of the particular programmable controller model in question. In addition, little information on practical applications was published. I wrote this book to aid in crossing these hurdles, for both the newcomer to the field and the experienced control engineer requiring a fresh perspective. It is intended to address both the quantitative and qualitative issues of programmable controllers. The most important thing that I can relate in this book is the application or "solution" power that programmable controllers bring to such a wide variety of control problems, and the actions that can be taken to solve those problems. If I can accomplish that, I shall consider having written this book worthwhile.

There are always those that aid in any process, for it is people, in the final analysis, who make any work possible through the understood phenomenon of collective action. Individuals that I would like to thank for their contributions to this work include Ken Jannotta and Bill Fountain for their help in exploring product structuring and training techniques; Rod Sipe for support in securing the many product photos used; and Gus Hof for technical systems aids. Special thanks are due to Dick Tanner for his insight on the strategic nature of the programmable controller marketplace, and to Ormand Austin for his invaluable reviews of the work.

Acknowledgments and thanks are also due to the many programmable controller manufacturers that made technical contributions to the development of this work. These are: Allen Bradley (Div. Rockwell), Automation Systems, Cincinnati Milacron, DeVilbiss, Eagle Signal, Eaton Corp.—Cutler Hammer, Eaton Leonard, Furnas Electric, General Electric, Giddings & Lewis, Gould Incorporated, Guardian, Honeywell, Industrial Data Terminals, McGill, Metra, Mitsubishi, Omron, Opto 22, Process & Instrumentation Design, Reliance Electric, Siemens Allis Automation, Square D, Struthers Dunn, Telemechanique, Texas Instruments, Toshiba, Triconex, Westinghouse.

This book is dedicated to Donna, Jessica, and Lindsey. Without their support, the work could not have been completed.

David G. Johnson

Contents

1
Introduction of
Programmable Controllers

*From a simple heritage, these remarkable systems have
evolved to not only replace electromechanical devices,
but to solve an ever-increasing array of control problems
in both process and nonprocess industries. By all
indications, these microprocessor powered giants will
continue to break new ground in the automated factory
into the 1990s.*

1.1 HISTORY

In the 1960s, electromechanical devices were the order of the day
as far as control was concerned. These devices, commonly known
as *relays*, were being used by the thousands to control many se-
quential-type manufacturing processes and stand-alone machines.
Many of these relays were in use in the transportation industry,
more specifically, the automotive industry. These relays, installed in panels and control cabinets (see Figure 1.1), used
hundreds of wires and their interconnections to effect a control
solution. The performance of a relay was basically reliable — at
least as a single device. But the common applications for relay
panels called for 300 to 500 or more relays, and the reliability
and maintenance issues associated with supporting these panels

Figure 1.1 Photo relay panel installed in cabinet. (Courtesy of General Electric.)

became a very great challenge. Cost became another issue, for in spite of the low cost of the relay itself, the installed cost of the panel could be quite high. The total cost including purchased parts, wiring, and installation labor, could range from \$30 to \$50 per relay. To make matters worse, the constantly changing needs of a process called for recurring modifications of a control panel. With relays, this was a costly prospect, as it was accomplished by a major rewiring effort on the panel. In addition, these changes were sometimes poorly documented, causing a second-shift maintenance nightmare months later. In light of this, it was not uncommon to discard an entire control panel in favor of a new one with the appropriate components wired in a manner suited for the new process. Add to this the unpredictable, and potentially high, cost of maintaining these systems as on high-volume motor vehicle production lines, and it became clear that something was needed to improve the control process — to make it more reliable, easier to troubleshoot, and more adaptable to changing control needs.

That something, in the late 1960s, was the first programmable controller. This first 'evolutionary' system was developed as a specific response to the needs of the major automotive manufacturers in the United States. These early controllers, or programmable logic controllers (PLC), represented the first systems that (1) could be used on the factory floor, (2) could have their 'logic' changed without extensive rewiring or component changes, and (3) were easy to diagnose and repair when problems occurred.

It is interesting to observe the progress that has been made in the past 15 years in the programmable controller area. The pioneer products of the late 1960s must have been confusing and frightening to a great number of people. For example, what happened to the hardwired and electromechanical devices that maintenance personnel were used to repairing with hand tools? They were replaced with 'computers' disguised as electronics designed to replace relays. Even the programming tools were designed to appear as relay equivalent presentations. We have the opportunity now to examine the promise, in retrospect, that the programmable controller brought to manufacturing.

1.2 BASIC CONCEPTS

All programmable controllers consist of the basic functional blocks shown in Figure 1.2. We'll examine each block to understand the relationship to the control system. First we look at the CPU, as it is the heart (or at least the brain) of the system. It consists of a microprocessor, logic memory for the storage of the actual control logic, storage or variable memory for use with data that will ordinarily change as a function of the control program execution, and a power supply to provide electrical power for the processor and memory. Next comes the I/O block. This function takes the control level signals from the CPU and converts them to voltage and current levels suitable for connection with factory grade sensors and actuators. The I/O type can range from digital

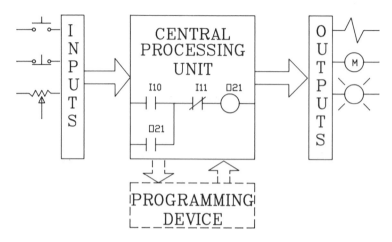

Figure 1.2 Diagram of basic programmable controller functions blocks.

(discrete or on/off), analog (continuously variable), or a variety of special purpose 'smart' I/O which are dedicated to a certain application task. The programmer is shown here, but it is normally used only to initially configure and program a system and is not required for the system to operate. It is also used in troubleshooting a system, and can prove to be a valuable tool in pinpointing the exact cause of a problem. The field devices shown here represent the various sensors and actuators connected to the I/O. These are the arms, legs, eyes, and ears of the system, including pushbuttons, limit switches, proximity switches, photosensors, thermocouples, RTDs, position sensing devices, and bar code readers as input; and pilot lights, display devices, motor starters, DC and AC drives, solenoids, and printers as outputs. Chapters 4 to 10 examine in detail the functions and contributions of the CPU, I/O and programming/documentation systems.

No single attempt could cover its rapidly changing scope, but three basic characteristics can be examined to help classify an industrial control device as a programmable controller.

1. Its basic internal operation is to solve logic from the beginning of memory to some specified stopping point, such as end of memory or end of program. Once the end is reached,

the operation begins again at the beginning of memory. This scanning process continues from the time power is supplied to the time it is removed.

2. The programming logic is a form of a relay ladder diagram, Normally open, normally closed contacts, and relay coils are used within a format utilizing a left and a right vertical rail. Power flow (symbolic positive electron flow) is used to determine which coils or outputs are energized or de-energized.

3. The machine is designed for the industrial environment from its basic concept; this protection is not added at a later date. The industrial environment includes unreliable AC power, high temperatures (0 to 60°C), extremes of humidity, vibrations, RF noise, and other similar parameters.

1.3 GENERAL APPLICATION AREAS

The programmable controller is used in a wide variety of control applications today, many of which were not economically possible just a few years ago. This is true for two general reasons: (1) their cost effectiveness (that is, the cost per I/O point) has improved dramatically with the falling prices of microprocessors and related components, and (2) the ability of the controller to solve complex computation and communication tasks has made it possible to use it where a dedicated computer was previously used.

Applications for programmable controllers can be categorized in a number of different ways, including general and industrial application categories. We will see more applications in a later chapter, but it is important to understand the framework in which controllers are presently understood and used so that the full scope of present and future evolution can be examined. It is through the power of applications that controllers can be seen in their full light. Industrial applications include many in both discrete manufacturing and process industries. Automotive industry applications, the genesis of the programmable controller, continue to provide the largest base of opportunity. Other industries, such as food processing and utilities, provide current development opportunities.

There are five general application areas in which programmable controllers are used. A typical installation will use one or more of these integrated to form a complete solution to the control system problem. The five general areas are explained briefly below.

Sequence Control. This is the largest and most common application for programmable controllers today, and is the closest to traditional relay control in its 'sequential' nature. Because of the very general nature of this category, it is sometimes difficult to understand the breadth of power that it brings to so many applications. From an applications standpoint, sequence control is found on individual machines or machine lines, on conveyor and packaging machinery, and even on modern elevator control systems.

Motion Control. This is the integration of linear or rotary motion control in the programmable controller. This could be a single or multiple axis drive system control, and can be used with servo, stepper, or hydraulic drives. In early systems, a stand-alone servo drive would be connected to the programmable controller with a series of individual conductors to discrete inputs and outputs. Newer systems integrate this functionality directly into the I/O racks through the use of special I/O boards dedicated to motion control. This eliminates the need to interface the two devices together with discrete I/O. Programmable controller motion control applications include an unending variety of machinery; metal cutting (grinders), metal forming (press brake), assembly machines, and multiple axes of motion can be coordinated for both discrete part and process industry applications. Examples of these would include cartesian robots, and many web related processes that is, film, rubber, and nonwoven textile systems.

Process Control. This is the ability of the programmable controller to control a number of physical parameters such as temperature, pressure, velocity, and flow. This involves the use of analog (continuously variable) I/O to achieve a closed-loop control system. The use of Proportional-Integral-Derivative, (PID) software allows the programmable controller to replace the function of stand-alone loop controllers. Another alternative, described later, is to integrate the loop controllers with the programmable controller, retaining the best features of each. Typical examples of applications include plastic injection molding machines, extrusion process machines, heat treat furnaces, and many other batch-type control applications.

Data Management. The ability to collect, analyze, and manipulate data has only become possible with programmable controllers in the last few years. With the advanced instruction sets and expanded variable memory capacities of the newer programmable controllers, it is now possible for the system to act as a data concentrator, collecting data about the machine or process it is controlling. This data can then be compared to reference data in the controllers memory, or can be sent via a communication function to another intelligent device for analysis or report generation. Any comments that are made about the importance and growing use of data management are probably understatements considering the leap-frog technical solution capabilities it brings to a wide range of applications. Data management is frequently found on large materials handling systems, in unmanned flexible manufacturing cells, and in many process industry applications, that is, paper, primary metals, and food processing.

Communications. This is the ability for the programmable controller to have a 'window' to other programmable controllers and intelligent devices. One of the most active development areas in today's industrial control arena and much Local Area Network (LAN) activity is currently driven by the MAP communications standard. The Manufacturing Automation Protocol (MAP), an activity initiated by General Motors, is intended to connect multivendor intelligent devices, including programmable controllers, into a coherent, efficient control network. In addition, higher performance control-oriented networks, sometimes referred to as subnets, offer the ability to tie together a small number of programmable controllers to form an 'island of automation.' Communications are most often used with a factory host computer, for the purpose of collecting process data and configuring the controllers for a certain production sequence. The majority of communication networks involving programmable controllers today are in the automotive industry. These are used in the production of a variety of engines, transmissions, and in assembly and paint operations. It is clear, however, that other industries will be catching up in the use of factory communications, most notably perhaps the aircraft, chemical, and heavy equipment industries.

As mentioned earlier, these applications may be found alone, or in combination on a variety of apparatus, Table 1.1 lists a sample of the many applications and industries using programmable controllers. Let's examine two hypothetical cases, the first a simple

Table 1.1 Programmable Controller Applications

Annunciators	Injection Molding
Auto Insertion	Assembly
Bagging	Motor Winding
Baking	Oil Fields
Blending	Painting
Boring	Palletizers
Brewing	Pipelines
Calendaring	Polishing
Casting	Reactors
Chemical Drilling	Robots
Color Mixing	Rolling
Compressors	Security Systems
Conveyors	Stretch Wrap
Cranes	Slitting
Crushing	Sorting
Cutting	Stackers
Digestors	Stitching
Drilling	Stack Precipitators
Electronic Testing	Threading
Elevators	Tire Building
Engine Test Stands	Traffic Control
Extrusion	Textile Machine
Forging	Turbines
Generators	Turning
Gluing	Weaving
Grinding	Web Handling
Heat Treating	Welding

sequencing application, and the second a machine which integrates all five application categories.

A stretch-wrap is a machine which automatically wraps plastic film or other material around and unitizes quantities of material (bags, boxes) in the most optimum way on a pallet for shipping. It normally involves sections of a powered conveyor and an operating area for the product layers to be placed together and wrapped. The programmable controller is used to sequence the operation of the stretch-wrap machine, sensing the presence of a pallet, and using actuators and other sensors to move that pallet automatically to its proper place in the machine. Stretch-wrap patterns are 'remembered' by the controller, and alternating wraps are staggered to improve the transportability of the loaded pallet. When a pallet is complete, it is sequenced by the controller to the exit conveyor section, to be loaded on a truck. A stretch-wrap machine is shown in Figure 1.3.

Plastic injection molding machines produce millions of items used in daily life. A programmable controller can be used here in a combination of the application areas listed above. Sequence control is used to arrange the various actions needed to load, unload, and execute the molding process. Motion control is used to control the velocity and position of the ram, which provides the force required to perform the actual molding process. Process control is used to control the temperature and pressure of the molding cavity. Data on the number and quality of parts produced is collected and concentrated using data management functions, and this data summary is made available to a communications network, for integration into a plant-wide network of molding machine controllers and factory host computers. Figure 1.4 shows a typical plastic injection molding machine.

As you can see from this introduction, applications for programmable controllers are extremely diverse and varied. This diversity will continue to increase in the future as cost-effectiveness and creativity combine to form an unbeatable team for automation productivity. In Chapter 11 we will examine in depth the solution power of these industrial strength microprocessor muscles—programmable controller applications.

Figure 1.3 Photo of stretch wrap machine. (Courtesy of Lantech Inc.)

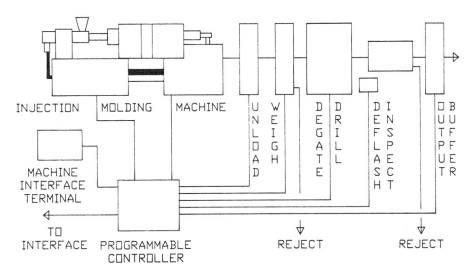

Figure 1.4 Diagram of plastic injection molding machine.

1.4 OPERATING ENVIRONMENT CONSIDERATIONS

One of the primary differences between programmable controllers and their general pupose computer cousins is the environment in which they can operate. A factory can provide some very un- friendly, sometimes hostile, conditions. All manner of dirt, grime, shock, vibration, temperature and humidity extremes, and electromagnetic field interference are present for the life of a typical programmable controller application. Couple this with an unreliable source of power, and you have the makings of a very difficult design challenge, for both the programmable controller and the installation process. We will examine briefly the various areas of concern for the programmable controller environment.

Dirt. This is just plain day-to-day factory dirt, and can be counted on for causing problems in many electrical installations of all types. All programmable controllers are enclosed to one degree or another, but the best insurance here is to mount the controller in a good quality cabinet, NEMA 12 or NEMA 4, and keep it closed as much as possible. The National Electrical Manufacturers Association

(NEMA) defines standards for cabinet design for a variety of environments, including oil- and water-tight. And while a commonsense cabinet protection may seem obvious, on two- and three-shift operations the cabinet doors may be left open for extended periods, especially during production pressured maintenance efforts. Most controllers can withstand an astonishing amount of abuse, but it makes practical sense to protect them as much as possible.

Shock and Vibration. Most controllers are designed and tested to substantial limits for shock and vibration. In turn, many applications deliver substantial levels of shock and vibration. Most designs will work well in the average factory environment, but if your installation calls for above-average expected levels of shock and vibration, you should consult the manufacturer for his test limits and experience. If these are not suitable for your application, you should inquire about the availability of special mounting techniques to reduce the shock and vibration impact. Special applications that may require vibration considerations include some nontranditional application areas such as mining, marine, and railroads.

Temperature Extremes. The industrial standard ambient operating temperature at the time of this writing is 0 to 60°C (or 32 to 140°F). Again, this is quite suitable for the majority of factory installations, but there are conditions that can cause trouble. At the high end, the internal temperature of a closed, nonventilated control cabinet can exceed 60°C in summer peaks, but may not cause a problem since the controllers have the traditional design tolerance band. If you experience erratic operation of the controller during extremely hot periods, supplemental cooling may be required. The low end of the temperature range can cause similar concerns. Normally the ambient heat generated from the operating controller will provide sufficient warmth for reliable operations. However, if low temperatures are suspected as the reason for erratic operations, supplemental heat may be required. This is especially true if the controller will be 'cold started,' since the advantage of self-generated ambient heat will not be consistently available. As a point of interest, some programmable controller manufacturers are seeing an increase in the call for lower temperature ratings for their controllers. This is especially true for those that operate on 24 volts DC. Some installations are being made in remote areas, where heat and AC line power are considered luxuries. Included here are remote oil pumping stations and railroad track switching applications.

Humidity. The standard here is 5 to 95% relative humidity, noncondensing. Note the 'noncondensing,' as this suggests that the controller will not survive in conditions where moisture would collect on an extended basis. This is quite understandable and normally does not cause any design or operational problems with a good cabinet choice.

Interference. This can be an elusive culprit, but can be dealt with for most applications. Without proper design, electrical interference will wreak havoc on an otherwise sound programmable controller installation. Unpredictable, and sometimes dangerous, I/O or logic status changes can occur in heavy interference environments. Some controllers, because of their designs, will fare better in this environment than others. A few guidelines will improve the chances for a successful application. Good cabinet design choice along with avoidance of any known interference source in the physical placement will help. In the wiring of the controller power and I/O, segregation of 115 volt and low voltage (i.e., 24 volt AC/DC) will minimize induction coupling. In extreme situations, external filtering may be required although optically coupled I/O and power supply designs will normally provide adequate protection.

Power. Nothing is so unreliable as the power supply in a typical factory environment. Fluctuations in voltage and availability are common. The quality of line service can be improved by using conditioning equipment such as isolation transformers and line conditioners. Highly critical applications may warrant an uninterruptible power supply (UPS), which provides clean power and battery backup during power failures. Most installations will not require this extreme action, since the power that supplies the controller also supplies the rest of the manufacturing process. In that case it is important to understand, and design for, the orderly shutdown and automatic power-up sequence of the controller. Some applications will call for a manual restart after power failure for safety reasons.

1.5 DEDICATED MICROPROCESSOR BASED SYSTEMS— A CONTRAST

One of the premises on which the programmable controller is based is the idea that it is a general purpose control product, capable of being configured, modified, and tailored to the application by the original equipment manufacturer (OEM) or user that purchased the controller. The hardware and software of such a system is such that it can be supported by a normal maintenance electrician.

There is a category of control product that is much more narrow and optimized for a particular control task. It uses a sometimes unique design, customized for the particular application by the manufacturer. The system is normally programmed in a programming language not familiar to the average control engineer and maintenance electrician (Assembly), which makes the support of the system difficult. The advantages of such a system are its cost-effectiveness for the given purpose, and sometimes its efficiency of execution. For example, certain automatic testing systems employ dedicated microprocessor based systems, sometimes in conjunction with a programmable controller. In this case the microprocessor system is used to collect large amounts of process data at high speeds, and the programmable controller executes the proper coordination with the rest of the manufacturing process. Another example of an application where dedicated controls are sometimes used is high-speed packaging where a general purpose system does not provide the throughput needs.

The distinction between the programmable controller and the microprocessor-based system is beginning to blur as controllers continue to gain data processing and communication control attributes while the microprocessor-based system takes on discrete I/O. This trend, combined with the desire of factory personnel to 'standardize' on a hardware and software vendor(s), will make the dedicated system unattractive in all but the most demanding applications. Chapter 13 examines the continued evolution and directions of the programmable control industry.

1.6   PERSONAL COMPUTER IMPLICATIONS

Relative newcomers to the industrial control arena, personal computers are finding more and more tasks suitable for their application.   Most of the current uses of personal computers are related to program development and documentation of programmable controllers and dedicated microprocessor-based systems.   Recent design enhancements have been made by certain manufacturers to allow the computer to survive in a more hostile environment that was previously conceivable (see Figure 1.5).   Improvements in operating temperature limits and filters to prevent foreign materials from being drawn into the system, are allowing a much higher level of 'industrialization' than the office-grade product would allow.   This trend is system design will allow a higher level of system product to evolve with much greater computing power, coupled with a greatly enhanced intelligent I/O system.   The result will be an inherently distributed and intelligent system, allowing system con-

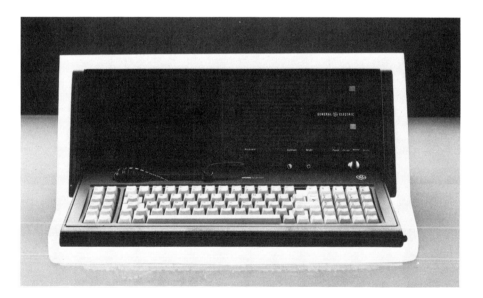

Figure 1.5 Photo of factory hardened personal computer. (Courtesy of General Electric.)

trol and communcations improvements not considered attainable
before. The favorable by-product of all this activity will be an
increase in productivity, with attractive payback and improved
corporate profitability.

There are two types of personal computer installations currently finding their way into industrial control applications. One is the
portable device, industrialized to survive on most factory floors,
at least for short periods of time. It is used for loading programs
into the programmable controller memory, and for on-line monitoring and troubleshooting. It is also normally used for program
development and documentation in an off-line mode. The second
personal computer installation type is a larger, more powerful system than the portable device. It will have more memory, both semiconductor and disk storage. These systems use a real-time multi-
tasking operating system, and are permanently installed in a standard

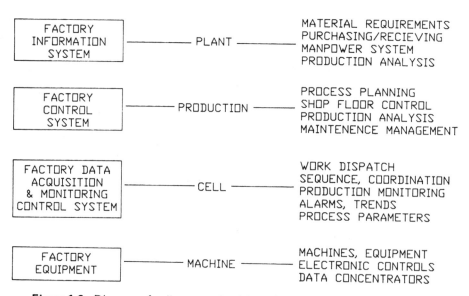

Figure 1.6 Diagram of cell automation hierarchy.

control cabinet using no fans for cooling. They are modular in design, allowing custom configurations and easy maintenance. They
are used as data concentrators in flexible manufacturing systems
at the cell or zone level. They are also used as system directors,
planning work scheduling in the cell along with the tools and materials required for a given sequence of operations. Figure 1.6
illustrates a typical hierarchal control scheme, showing the relationship of these 'cell controller' computers to programmable controllers and other intelligent devices.

1.7   FACTORY AUTOMATION
AND PROGRAMMABLE CONTROLLERS

Programmable controllers play a fundamental role in factory automation efforts. Their breadth of application capabilities allow
firms to use the products of one or two manufacturers in a wide
array of process and nonprocess areas. There are several levels of
evolution involved in factory automation advances in most plants.
Most firms will be interested in modernizing existing facilities
because the construction of entirely new facilities will not be economically feasible. In many instances, the first stage of automation
involves small contiguous production areas, concentrating on improving productivity in that area alone. It is not uncommon to
fund automation efforts through the savings of reduced in-process
inventory alone. These small contiguous areas have come to be
called 'islands of automation.' Progammable controllers are used
here to operate machinery within the island, including stand-alone
machine tools, dedicated machinery (i.e., packaging machinery),
and material handling equipment for both within the island and to
bring items in and out of the area.

 The normal extension of this type of automation is to tie
multiple islands together, both physically with bulk or unit handling systems, and through the data communication links of a proprietary or standard local area network. This control network,
linking programmable controllers and other factory automation
control products, allows real-time flexible execution of the

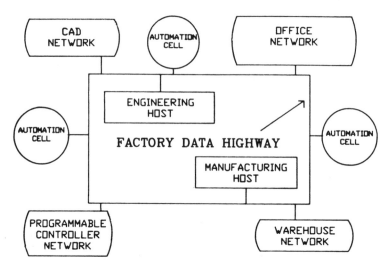

Figure 1.7   Diagram of factory automation system.

optimum production process. Figure 1.7 shows a factory automation system including programmable controllers, numerical controllers, robot and vision systems, and communication networks.

2
Logic Concepts

*Whenever the nature of the subject permits the
reasoning process to be without danger carried on
mechanically, the language should be constructed
on as mechanical principles as possible...*

George Boole

2.1 BINARY LOGIC—THE ON/OFF CONCEPT

In control systems found using programmable controllers, the pre-
ponderance of sensors and actuators are concerned with just two
states: ON or OFF. In logic circuits, this corresponds to true/
false, or in symbols, A and A'. To relate this to control sytems,
an input to the programmable control system from a pushbutton
has but two states: ON or OFF. That is, either the button is
being pushed, or it isn't. Once the state or change of state is
registered in the programmable controller memory, it can be used
in any number of ways, but the principle is fundamental. This is
the basic principle of binary logic, binary meaning consisting of
two things or parts. The binary numbering system is based on the
number two, and hence all sets are represented by combinations
of 0 and 1. For example, while it is clear that 00 in binary is 0,

and 01 is 1, it may not be clear that 10 equals 2, and 11 equals 3. We will examine these extensions of the binary numbering systems in details in Chapter 3. For now we will concern ourselves with the convention only that the 0 and 1 of binary logic correspond to the OFF and ON of control circuit inputs and outputs.

   To see this relationship more clearly, consider Figure 2.1(a). Here we have a simple control circuit consisting of two components: a switch A (perhaps from a pushbutton) and a load M. Assume that a sufficient voltage exists across the two vertical lines to power the load M. It should be clear that when the contacts on switch A change state and close, current will pass through it to the load M, thereby energizing that load. This is the fundamental binary variable case and illustrates the identity function A = M. Figure 2.1(b) shows the Venn Diagram for this same identity. The rectangle encloses all propositions under consideration and is referred to as the universal set. The circle represents the different existing conditions. For this simple case, we consider only two cases: (1) those for which the switch A is on (inside the circle), and (2) those for which the switch A is off (out-

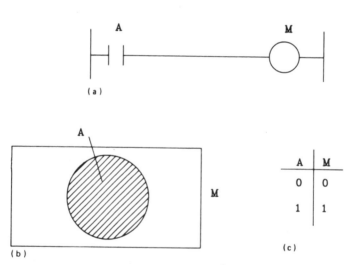

Figure 2.1 Diagram of logic IDENTITY function.

side the circle). The shaded area in side the circle indicates that the proposition under consideration (that is, M) is present if we are inside the circle. Figure 2.1(c) shows the corresponding truth table for the function of A = M.

2.2 COMBINATIONAL LOGIC—AND, OR, NOT

The fundamentals illustrated in the previous section can be extended and combined to perform more complex logic functions. Here we will examine the AND, OR, and NOT functions. These functions, when combined, can execute what is referred to as combinational logic.

Let's examine the AND function. Referring to the illustrations in Figure 2.2 you can see that inputs A and B in different combinations of ON and OFF (0 and 1) produce a different output. When both A AND B are present, (i.e., 1), the output is 1, or ON. Any other combination results in the output being OFF. In the electrical circuit shown in Figure 2.2(a), the AND function is illustrated by two series contacts. Therefore when both contact A AND contact B are closed, the load M is energized

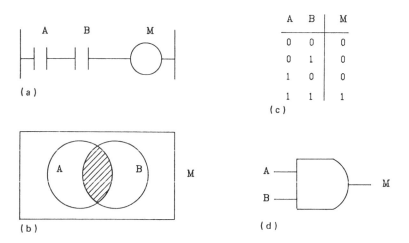

Figure 2.2 Diagram of logic AND function.

or ON. The truth table and the Venn diagram show this arrangement. The area of the circles A and B which overlap represent the condition when M is true or 1. Figure 2.2(d) shows the common symbol for the AND function, also called the AND gate.

Figure 2.3 shows the relationship for the OR function. Here, if either A or B is present, the function is true, and the output M is energized. The Venn diagram shows that anywhere within the area of circle A or B, including the overlapping areas, M is true. It is clear from the corresponding electrical diagram (Figure 2.3 (a)) that when contact A OR contact B is closed, the load M will be energized. Figure 2.3(d) illustrates the symbol for the OR function.

To utilize combinational logic fully, another convention is needed in logic representations. Figure 2.4 describes the NOT function. This is simply the complement, or opposite, of a given reference. The truth table shows this as "whenever A is present, M is NOT." In the electrical system, this is shown by a normally closed contact. It follows that whenever A is NOT executed, current flows to load M, and when A changes state to open, current

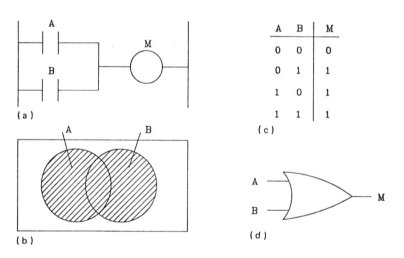

Figure 2.3 Diagram of logic OR function.

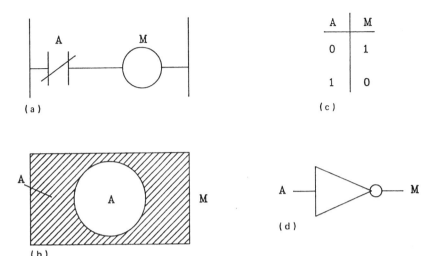

A	M
0	1
1	0

(c)

(a)

(b)

(d)

Figure 2.4 Diagram of logic NOT function.

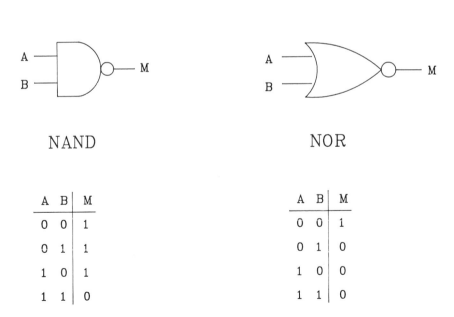

NAND

NOR

A	B	M
0	0	1
0	1	1
1	0	1
1	1	0

A	B	M
0	0	1
0	1	0
1	0	0
1	1	0

Figure 2.5 Diagram of NAND and NOR functions.

ceases to flow into load M. The Venn diagram shows this as A'
being the normally closed equivalent, and being present for the
"universe" M as a shaded area. The NOT function is used many
times in combination with the AND and OR functions to complement the output. In this case the resulting logic functions are
called NAND and NOR functions, respectively (see Figure 2.5).
Can you think of the electrical diagrams that correspond to the
two functions?

2.3 SEQUENTIAL LOGIC—FLIP-FLOPS, TIMERS, AND COUNTERS

So far, we have been concentracting on logic concepts and circuits
with the assumption that time was not a factor in the control execution. In reality, this is probably safe with combinational logic
circuits, although there is a finite propagation time involved. This
is ususally in the nanosecond range and is typical of that found in
the solid state, so-called "hardwired" logic. With a minimum of
design care, this propagation phenomenon will not affect a predictable network output.

Most real-world control circuits, however, contain an element
of time for successful control execution. Examples include a time-
delayed startup of a synchronized bulk handling conveyor system,
and machinery using time-based control to sequence individual
operations. This category of logic circuit is referred to as sequential logic and involves the use of static storage of logic status,
time delay, and discrete counting. We will examine each below.
Figure 2.6 illustrates each in both solid state and control versions.

Static storage of logic status is the "capturing," in essence, of
a transition of some selected logic status. The solid state device
for this function is a flip-flop. It is arranged so that a logic transition from 0 to 1 on one of the device's inputs causes a corresponding transition in the output line. The new status then becomes
stored statically in the device until another 0 to 1 transition is
detected.

The timer function is straightforward. The device is enabled
aster being assigned a preset time. After the time in question has

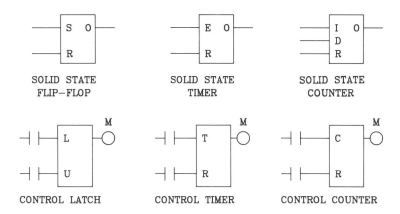

Figure 2.6 Diagram of sequential logic functions.

elapsed, the output transitions. The transition can either be time-on or time-off, depending on the control circuit design called for. A reset line allows the device to begin timing again at zero.

Counters are equally straightforward with one additional feature-flexibility. They are tasked with maintaining the current cumulative count, so they must be able to sense both increments and decrements in count input. Otherwise they have the same output and reset options.

2.4 BOOLEAN ALGEBRA

The combinational and sequential logic functions we have examined above are only useful when arranged in a manner to solve problems. This section illustrates two such arrangements and will review basic boolean algebra. If additional depth is required, please refer to any of a number of good texts on boolean algebra representations and minimizations.

First we look at an arrangement of combinational gates only. Later this will be expanded to include sequential functions. Figure 2.7 shows two simple examples. In 2.7(a) the gates are arranged so that either A and B, or C and D, must be present, or

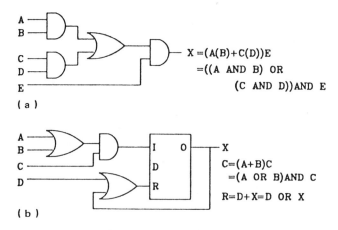

Figure 2.7 Diagram of combinational and sequential example.

true, for X to be active. In addition, E must always be present, or true for X to be active. In a practical situation, E might be a keyswitch enable circuit, and A, B, C, and D are pairs of enable circuits, perhaps from two separate physical locations. X might be the start circuit for a compressor, with demand for compressed air coming from two places within the plant.

Figure 2.7(b) shows both combinational and sequential logic. This could be part of the logic for a packaging machine, with C being the machine enable for counting items to be packaged. The counter would have a preset number to count up to, and at that time the output X would enable. A and B would represent sensor inputs sensing presence of two different sizes of items to be packaged. D could represent a manual reset input. With C present, A or B present would increment the counter. When the accumulated count reached the preset, the output X would be active. This would, among other things, reset the counter to zero to allow another counting process to begin.

These two examples show some of the practical examples possible using both combinational and sequential logic. In the next section we will see the extension of this concept to industrial control applications.

2.5 PROGRAMMABLE CONTROLLERS—A TRANSITION

So far, we have seen logic circuits only in their static, gate-or relay-based forms. Our purpose is, of course, to move on to programmable controllers, and in this section we will do so. The first task is to present the significant advantages of making the transition, and then we will examine the various conversion examples and symbologies used in programmable controller practice today.

In the hardwired systems typical of relay-and logic-gate-based systems today, flexibility is nonexistent. Flexibility is the key advantage to using programmable controllers over these earlier systems. As illustrated in Chapter 1 dedicated systems, both microprocessor and non-microprocessor based, have some narrow cost and execution efficiency advantages, but today's general purpose programmable controllers are almost always better suited for a vast array of industrial control applications. Flexibility comes from the key word—programmable. Today's controllers execute a control program from stored memory; a program memory that can be changed dynamically to suit the changing application by those people that use the system, not some distant, mysterious "computer guru." The secret to accomplishing this flexibility is the programming language used to accomplish the control program creation. It is very similar to the relay circuit symbology used for the past 40 years by the majority of plant floor electricians in the United States. It is based on the concept of power flow, and uses contacts and (relay) coils as its foundation items. There are differences, but most of these can be highlighted and overcome by a short training course from the programmable controller manufacturer. Even a course is not always necessary, providing the manufacturer has provided comprehensive documentation of the hardware and programming techniques.

Table 2.1 shows some of the common sensors and actuators that represent the physical inputs and outputs of the programmable control system. Included are the pushbutton and limit switches, solenoids, and pilot lights. These are the arms, legs, eyes, and ears of the control system, and their status is registered and controlled by the software of the control program created specifically for the application at hand. Unlike the hardwired system where multiple uses of coil contacts meant multiple contacts,

Table 2.1 Sensors and Actuators

Sensors	Actuators
Pushbuttons	Motor Starters
Limit Switches	Indicator Lights
Photo Cells	Variable Drives
Proximity Switches	Solenoids
Thermocouples	Control Relays
Potentiometers	Analog Valves
Load Cells	Chart Recorders

representations of this physical I/O can be used multiple times, limited only by the program memory capacity.

The transition from rigid circuits to programmable ones is illustrated in Table 2.2 and Figure 2.8. Table 2.2 shows contact and coil conventions and touches on a significant difference is

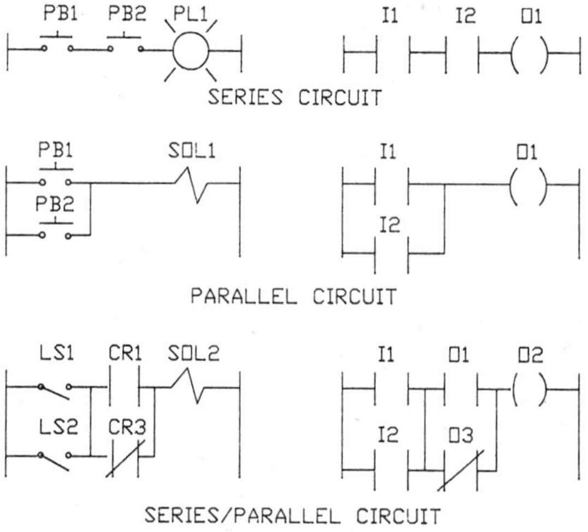

Figure 2.8 Diagram of conversion examples; relay to programmable control.

Table 2.2 Contact, Coil, and Mnemonic Conventions and Notations

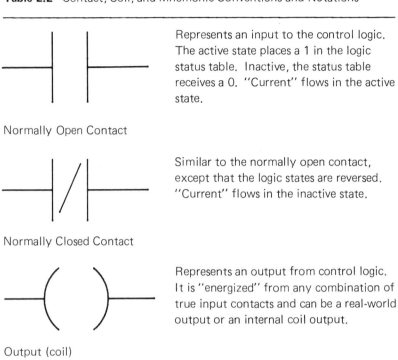

Normally Open Contact	Represents an input to the control logic. The active state places a 1 in the logic status table. Inactive, the status table receives a 0. "Current" flows in the active state.
Normally Closed Contact	Similar to the normally open contact, except that the logic states are reversed. "Current" flows in the inactive state.
Output (coil)	Represents an output from control logic. It is "energized" from any combination of true input contacts and can be a real-world output or an internal coil output.

using programmable controllers, the use of mnemonic functions in relay ladder logic to manipulate binary data. This capability, first introduced in the late 1970s, is allowing controllers to accomplish broader and more complex tasks than was ever thought possible. The ability to collect, manage, and concentrate/communicate data is one of the fundamental building blocks of today's factory automation tasks. Figure 2.8 highlights several examples of conversion of relay circuits to programmable controller circuits. How the controller executes control using this program will be examined in Chapters 4 and 5, the CPU and I/O system, respectively.

3
Numbering Systems and Coding Techniques

A fella' can no more tell you what he don't know than he can be from where he ain't been.

Origin Unknown

3.1 NUMBERING SYSTEMS AND CONVERSIONS—BINARY, OCTAL, DECIMAL, AND HEXADECIMAL

Numbers and numbering systems are used in many everyday occurrences to state, restate, or communicate some type of information. All programmable controllers, from the smallest to the largest and most complex, rely on efficient numbering systems to accomplish their tasks. While it is true that the internal working system of the microprocessor-based programmable controller uses the binary numbering system (that is 1 and 0), many parts of the rest of the system use other numbering systems. Examples include: (1) Input/Output systems that are addressed in octal, (2) displays that use the Binary Coded Decimal or BCD system, (3) register values that are displayed and used in decimal, and (4) documentation printouts that are referenced in hexadecimal format.

In this chapter we shall examine each type of system, present conversion methods from one to the other, and methods in which they are used with programmable controllers.

The first numbering system we shall examine is the binary system. It uses as its base the number 2, which means that only 1 and 0 are allowed for use in this system. A base for a numbering system refers to its reference—contrast this to decimal which uses as its base the number 10. Binary is the primary system used internally in the programmable controller, and is used extensively in most of the manufacturers' documentation, especially in the register and I/O formatting sections, and the diagnostics sections. Table 3.1 shows an example of how the binary system works. Using four digits, different combinations of 1 and 0 are used to represent the numbers 0 to 15 in decimal. You will note that the digit in the binary number that is the rightmost of the four changes the most frequently in the transition from 0 to 15. This is called the Least Significant Bit, or LSB. The digit that is leftmost changes the least frequently, and is referred to as the Most Significant Bit, or MSB. The notation of least and most significant positions in a number will be common in all of the numbering systems that we examine. If you think back on the familiar decimal system, you will recognize that this is true, but becomes second nature with use. Figure 3.1 illustrates a 10-bit binary number and its decimal equivalent. Note the procedure of starting with the least significant bit (or digit), multiplying it by the base, in this case 2, raised to the 0 power. This process continues using the next most significant bit and the $1st$ power, and so on, until the most significant bit has been processed. The results of these individual multiplications are added together, with the result being the desired conversion from binary to decimal. This procedure is the same for all number conversions. Figure 3.2 shows the common format for a word of memory in a programmable controller. This particular diagram uses a 16-bit format or word, and consists of two 8-bit bytes. There are still some limited models of controllers that use an 8-bit word format, but we will focus on the much more common 16-bit. This diagram also shows the positions of the least and most significant bits in the word.

The next system we will consider is the octal numbering system. As the name suggests, this system has as its base the number 8. The octal system is occasionally used as the I/O addressing convention in some programmable controllers, and as the data referencing convention in a few, especially those with eight-bit word system architectures. The octal system uses the numbers 0

Table 3.1 Decimal and Binary Equivalents

Decimal	Binary
0	0000
1	0001
2	0010
3	0011
4	0100
5	0101
6	0110
7	0111
8	1000
9	1001
10	1010
11	1011
12	1100
13	1101
14	1110
15	1111

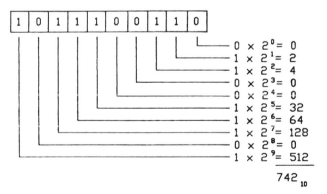

$$0 \times 2^0 = 0$$
$$1 \times 2^1 = 2$$
$$1 \times 2^2 = 4$$
$$0 \times 2^3 = 0$$
$$0 \times 2^4 = 0$$
$$1 \times 2^5 = 32$$
$$1 \times 2^6 = 64$$
$$1 \times 2^7 = 128$$
$$0 \times 2^8 = 0$$
$$1 \times 2^9 = 512$$

$$742_{10}$$

Figure 3.1 Diagram of binary to decimal conversions.

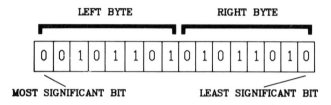

Figure 3.2 Diagram of sixteen-bit word format.

through 7 as its operating range. There are no 8 or 9, and the progression in counting follows a "six, seven, ten, eleven. . ." sequence. This is shown in Table 3.2. The process of converting the base eight numbers to decimal or base ten numbers, is similar to that shown for binary numbers and an example is shown in Figure 3.3.

The last system we shall consider for now, is the hexadecimal. This numbering system is based on the number 16. The name comes from hex for six and deci for ten; or six plus ten. Hexadecimal numbers are used a great deal in many programmable controller systems. They are especially useful as references on the programming device, as a great amount of binary information can be encoded in a condensed hexadecimal format. This is illustrated in Table 3.3, contrasting hexadecimal with both decimal and binary. You will note that only one character of information in the hexadecimal format is required to encode four characters of binary information. Figure 3.4 shows a conversion example of hexadecimal to decimal and hexadecimal to binary.

3.2 CODING TECHNIQUES AND CONVENTIONS

In many applications involving programmable controllers, it is necessary to encode and decode information. This normally involves interaction with external devices like CRT terminals or thumbwheel switch interfaces. These external devices communicate to the programmable controller primarily in one of two ways. One is to use regular input/output lines of a certain voltage, usually 24 volts DC. This is most commonly used with devices like thumbwheel switch interfaces and digital LED display devices.

Table 3.2 Decimal and Octal Equivalents

Decimal	Octal
0	0
1	1
2	2
3	3
4	4
5	5
6	6
7	7
8	10
9	11
10	12
11	13
12	14
13	15
14	16
15	17

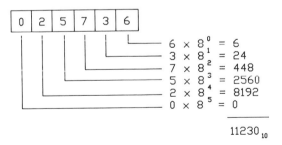

Figure 3.3 Diagram of octal to decimal conversions.

Table 3.3 Hexadecimal Equivalents in Binary and Decimal

Binary	Decimal	Hexadecimal
0000	0	0
0001	1	1
0010	2	2
0011	3	3
0100	4	4
0101	5	5
0110	6	6
0111	7	7
1000	8	8
1001	9	9
1010	10	A
1011	11	B
1100	12	C
1101	13	D
1110	14	E
1111	15	F

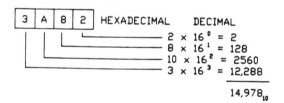

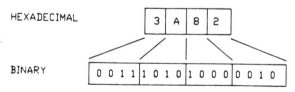

Figure 3.4 Diagram of hexadecimal to decimal and binary conversions.

The second is to use digitally encoded information sent and received through one of the programmable controller's data communication interfaces. We will examine some of the more common techniques of coding below along with some system examples.

The fundamental technique used in this effort is binary encoding. As we saw in Chapter 2, the binary numbering system consists of using the numbers 0 and 1 in combinations required to represent the various decimal numbers needed. This is what all digital computers ultimately use for their raw computational tasks. There are several codes that are used to represent letters, numbers, and symbols. The three most common are listed below, and have been established as industry standards.

ASCII
GRAY
BCD

ASCII is an acronym for the American Standard Code for Information Interchange and is pronounced "ask-ee." This code is used for the communication for both letters and numbers, and is commonly found in use with a peripheral device like a CRT or printer. The ASCII code can be 6-, 7-, or 8-bits of length for each number, letter, or symbol. Even though it may appear that there are only 26 letters and the decimal numbers 0 through 9 to represent, there are actually many more. We must accommodate the upper and lower case of each letter, and also provide for the encoding of a number of control characters used in setting up communication with the external device. In spite of this newly expanded task, seven of the eight bits in each of our defined bytes are enough to delineate our code requirement. We can see this by recalling that 2 to the the $7th$ power equals 128. The eighth bit in the byte is sometimes used to define communication parity for error checking purposes. Note that two 8-bit byte encoded ASCII representations can be compacted in one 16-bit (two byte) word of the programmable controller memory. A partial ASCII table is shown in Table 3.4, a complete table is included as Appendix C.

The Gray code is a cyclic code used primarily with position sensing devices such as digital encoders. This code system is normally part of the inner workings of the encoder system, and is,

Table 3.4 Partial Table of ASCII Values

Octal	Parity	Hex	ASCII Character
101	EVEN	41	A
102	EVEN	42	B
103	ODD	43	C
104	EVEN	44	D
105	ODD	45	E
106	ODD	46	F
107	EVEN	47	G
110	EVEN	48	H
111	ODD	49	I
112	ODD	4A	J
113	EVEN	4B	K
114	ODD	4C	L

therefore, transparent to the user of such apparatus. A Gray code table is shown in Table 3.5. It was developed for use with high-speed sensing systems, and only allows one bit of information to change for each increment or decrement. This provides for a more accurate and predictable sensing/decoding system. In today's advanced programmable controller systems, many have as available options an intelligent input/output module for use with an incremental position encoder. All of the position sensing, data latching, and accuracy maintenance is done in the intelligent module, relieving the user of the task of interfacing the encoder, and writing the program to decode the position data.

Binary Coded Decimal (BCD) is used to convert the binary that a machine uses to a binary that corresponds to the decimal system understoody by humans. Table 3.6 contrasts the BCD coding system to the pure binary and decimal numbering systems. Each of the numbers 0 through 9 are represented by a 4-bit binary number. Also illustrated in Table 3.6 is the ease of converting a multiple digit decimal number to its equivalent in the BCD system. The most common use of the BCD system in program-

Table 3.5 Gray Code Table

Binary	Gray code
0000	0000
0001	0001
0010	0011
0011	0010
0100	0110
0101	0111
0110	0101
0111	0100
1000	1100
1001	1101
1010	1111
1011	1110
1100	1010
1101	1011
1110	1001
1111	1000

Table 3.6 BCD Code, with Binary and Decimal Conversion Examples

BCD	Decimal	Binary
0000	0	0000
0001	1	0001
0010	2	0010
0011	3	0011
0100	4	0100
0101	5	0101
0110	6	0110
0111	7	0111
1000	8	1000
1001	9	1001

mable controllers is with peripheral input and output devices. Figure 3.5 shows a typical input from a thumbwheel interface, and a complementing seven-segment decimal LED display. The BCD is being supplied to, or received from, the programmable controller. The conversion from decimal to BCD and from BCD to seven-segment display is performed inside the respective peripheral device. Once inside the programmable controller the data is normally converted to binary. Each decimal digit requires four lines of input/output.

3.3 RELATION TO CONTROL APPLICATIONS

As we saw before, all of the information inside the programmable controller is stored in a binary format. For convenience, the information of a variable nature is stored in the register memory. Variable memory includes those types which change frequently, such as timer and counter accumulation data, binary encoded ASCII information, and mathematical references. Registers in most

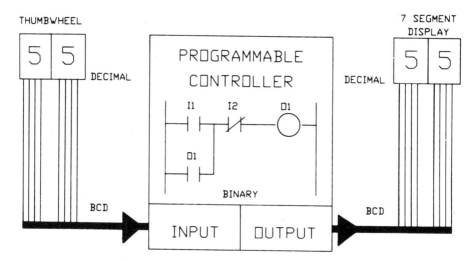

Figure 3.5 Diagram of thumbwheel in-LED out example of BCD conversion application.

modern programmable controllers use a 16-bit word format, and hence registers are 16 bits in length. The information stored in the variable memory can be formatted as binary, BCD, ASCII, or any other data form that can be configured in 0 and 1. Figure 3.6 shows two examples of this register format. You will note that a 16 bit register can store numbers from 0 to 65,535, if it is used in binary format. Occasionally a programmable controller will be capable of performing mathematical operations in signed arithmetic. In this case, the left-most, or most significant, bit will be used as a sign bit, changing the binary range to −32,767 to +32,767. A 0 in this bit location usually represents a positive number, and a 1 indicates a negative number.

A BCD format, on the other hand, used the 16-bit register in a different way. You can see in Figure 3.5 that it segments the register into four 4-bit parts, each segment representing a decimal digit. It is clear, therefore, that a register so formatted can only range from 0 to 9999.

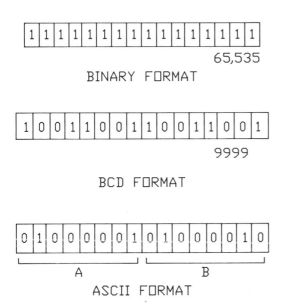

Figure 3.6 Diagram of register formats.

A simple example is the following application of information encoding and decoding. An automatic packaging machine will normally allow for an operator to adjust the prameters of the packaging process and observe the progress and status of the packaging process from an operator's panel which is mounted on the machine and connected to the programmable controller that is actually controlling the machine. The method by which the operator might adjust the packaging parameters is a thumbwheel input device. These devices are configured in single or multiple digit arrangements, and the decimal numbers representing the switch position are translated into BCD format in the thumbwheel switch assembly. This BCD-encoded information is then made available to the programmable controller in the form of input lines, four lines per decimal digit. The thumbwheel information is converted from BCD to binary and can then be used to change the preset value of a timer or counter in the programmable controller logic. Complementing this input of information might be an output of information to a LED display showing the machine status or other required information. In this case the binary information is converted into BCD by the programmable controller logic and is outputted to the LED display again using four lines per decimal digit. This example is illustrated in Figure 3.6.

4
The Central Processing Unit

Although referred to as the brain of the system, the Central Processing Unit in a normal installation is the unsung hero, buried in a control cabinet, all but forgotten.

4.1 BASIC FUNCTIONALITY

In a programmable controller system, the central processing unit (CPU) provides both the heart and the brain required for successful and timely control execution. It rapidly and efficiently scans all of the system inputs, examines and solves the application logic, and updates all of the system outputs. In addition, it also gives itself a checkup each scan to ensure that its structure is still intact. In this chapter we will examine the central processing unit as it relates to the entire system. Included will be the various functional blocks in the CPU, typical scan techniques, I/O interface and memory uses, power supplies, and system diagnostics.

4.2 TYPICAL FUNCTION BLOCK INTERACTIONS

In practice, the central processing unit can vary in its architecture, but consists of the basic building block structure illustrated in Fig-

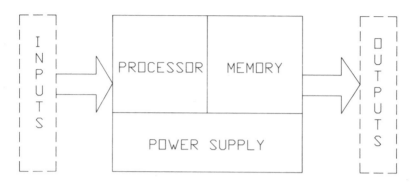

Figure 4.1 CPU block diagram.

ure 4.1. The processor section consists of one or more microprocessors and their associated circuitry. While it is true that some of
the older generation programmable controllers were designed without the luxury of using microprocessors, most modern systems use
either a single microprocessor such as the 8086 or Z-80, or multiple microprocessors such as the AMD2903, used in a bit slice
architecture. This multi-tasking approach is used in the multiple
microprocessor system to break the control system tasks into
many small components which can be executed in parallel. The
result of this approach is to achieve execution speeds that are orders of magnitude faster than their single-tasking counterparts. In
addition to efficiently processing direct I/O control information
and being programmable, the real advantage that microprocessor-
based systems have over their hardwired relay counterparts is the
ability to acquire and manipulate numerical data easily. It is this
attribute that makes programmable controllers the powerhouses
that they are today in solving tough factory automation problems.
The factory of tomorrow will run efficiently only if quality information about process needs and status of the process equipment
are known on a real-time basis. This can and will come about only
if the unit level controllers, including programmable controllers,
are empowered with the ability to collect, analyze, concentrate,
and deliver data about the process. As the market continues to
exhibit this demand, manufacturers are likely to continue to out-
fit their controllers with more and more variable memory, and
enhanced instruction sets to perform these tasks.

The memory segment shown in Figure 4.1 refers to the programmable controller's active storage medium. This can be either volatile or nonvolatile in design, and can be configured and used in a variety of ways for both executive program storage, with which the system executes its instructions, and application program storage, for the actual control program. More on this later in the chapter.

The power supply shown here is used for providing sufficient electrical current for the various semiconductors and other power-consuming devices on one or more of the CPU circuit boards. It can be arranged in a number of different physical ways. It may be located in the same chassis in which the CPU boards are located, or can be mounted in a stand-alone fashion, connected externally to the CPU chassis. Depending on the particular manufacturer's configuration, it may also provide power for some of the I/O functions, as well as the CPU.

4.3 SCAN TECHNIQUES

By definition and design, the programmable controller is dedicated to the continuous, repetitive task of examining the system inputs, solving the current control logic, and updating the system outputs. This task is referred to as scanning (sometimes called sweeping), and is accomplished in slightly different ways in each manufacturer's programmable controller. Since many of the variations are not material to the basic functionality of the system, we will only examine the basic varieties.

Figure 4.2 shows the functional operation of a typical scan mechanism. You'll notice that the I/O servicing is at the end of the scan cycle, and is also an integral part of the scan timing. This type of scan is referred to as synchronous scan and is used with very fast machines that can update all of the I/O without lengthening the scan time materially. A typical scan time in a modern programmable controller ranges from 10 to 100 milliseconds. Most controllers have a mechanism, a watchdog timer, to measure the scan length each cycle and initiate a critical alarm if the scan time exceeds a certain preset length, normally 150 to 200 milliseconds. Referring to Figure 4.1 again, the synchronous scan contains four other activities in addition to the I/O scan. Housekeep-

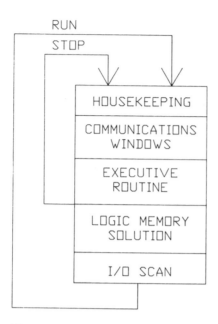

Figure 4.2 Scan sequence diagram.

ing refers to a small number of routine chores performed by the programmable controller to ensure that its internal structure is still healthy and functioning properly. Next comes the communications windows to allow structured communications to other devices in the system, or externally. Included in this group would be the programming device, special microprocessor-based communications modules to allow ultimate communication of the programmable controller system to another intelligent device. Next in line comes the executive routine, in which the actual base intelligence of the system is used to interpret the current control program. This interpretation is then used in the next step to solve the current control logic program. The last step of this basic scan process is to integrate the currently interpreted control logic program with the most current input statuses from the I/O scan, and to update the output statuses with the current results.

 The primary variation of this basic scanning technique comes from architectures that service and update I/O with a separate processor, asynchronous to the main logic solution scan. This alternative is common in systems where serial communication is used to control and update racks of remotely mounted I/O. It is also used where all of the I/O is serial, and run in multiple channels, to suit a particular systemconfiguration need. This parallel asynchronous scanning technique has the advantage that it allows extensive flexibility in configuring a programmable controller system for a particular application need. It has the disadvantage that while the basic scan rate maybe fast enough to suit an application, the I/O scan(s) may actually be longer than the primary CPU scan. This can cause problems in a fast acting system in that the logic solution can occur with relatively "old" input data from the remote I/O channel. While this is at times bothersome, the more dramatic case involves a peculiarity of some programmable controllers in that they may allow input and output data to be updated on separate time bases, providing the possibility of "bad" logic solutions and unpredictable machine actions.

As part of the basic CPU structure, a number of error checking procedures are used to maintain a high level of integrity in the communications between itself and its subsystems. This can involve both the internal subsystems, such as the memory, and the so-called external subsystems, for example the I/O system. The more common error-checking schemes are outlined below. The first and most common is *parity*. This is used on many communications link subsystems to detect errors by examining the number of "ones" in each byte of information received, and comparing the total number in any one byte to a predetermined choice of even or odd parity. This corresponds to the total of ones in the byte summing to an even or odd number. This has the disadvantage of being able to detect a single-bit error, that is where a zero or one has changed state during some operation; but cannot detect two single but opposing bit changes in a byte of memory that cancel each other out and still result in the correct parity. Checksum, the second most commonly used method for error checking, involves the examination of a block of memory for errors as compared to an individual word as done in parity checking. The procedure involves the adding of a single word of memory to a block that is unique to that block. Common varieties of the checksum are the Cyclic Redundancy Check (CRC) and the Longitudinal Redundancy check (LRC). The checksum advantage over the parity check is that it more efficiently uses memory. The third error checking method that we will consider is Error Detection and Correction (EDC). It is used in the more sophisticated programmable controllers provided by a few manufacturers today. In essence it involves a number of complex error correcting codes implemented in the hardware. The Error Detecting and Correcting method has the added advantage that it can sense and correct single-bit errors, while only sensing double bit errors.

4.4 I/O CONTROL

Today's modern programmable controller includes a sophisticated method to control the CPU's execution of the Input/Output chain. This is referred to as I/O control, or sometimes Bus control. This is actually handled in different ways, depending on the type and style of controller involved. In the small programmable controller (such as shown in Figure 4.3), the I/O servicing is performed as an integral part of the primary microprocessor used to control all of the major functions. In medium- and large-sized systems, it is common to include a microprocessor board or subsystem to handle the execution of the I/O updating. This is especially important in the systems that update I/O separately, or asynchronously from the main scan.

Regardless of the way it is achieved, the I/O control, or updating, is performed for the same reasons. For a successful scan sequence, an accurate execution of the signal level communications to the physical Input and Output modules is required. It is then, and only then, that any changes in the I/O status can be physically updated to the actuators or from the sensors. Later in Chapter 5 we will examine more about how the Input and Output modules are used.

4.5 MEMORY—USES AND STRUCTURE

Figure 4.1 shows how the programmable controller memory relates to the other functional blocks in the CPU. It is memory, along with a microprocessor to exercise it, that separates today's programmable controller from its predecessor. Current advances

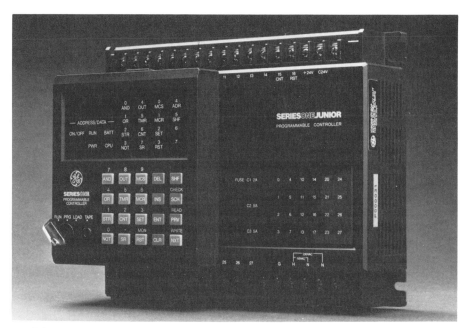

(a)

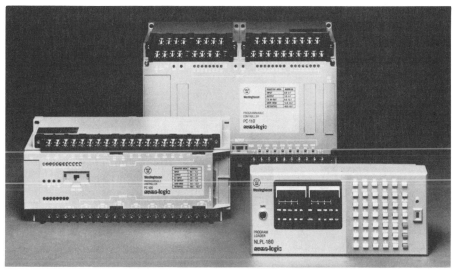

(b)

Figure 4.3 Examples of small programmable controllers. (a) Courtesy of General Electric; (b) courtesy of Westinghouse.

in memory allows both the rapid creation and efficient editing of control programs used to run the manufacturing processes. Different types of memory are used in a variety of programmable controllers for different application or design reasons. Let's examine some of them in detail.

There are two basic memory categories used by programmable controllers, or for that matter, any microprocessor-based system. They are volatile and nonvolatile. Volatile means that the contents of the memory have no means to remain intact without an external power source connected to maintain the data integrity. Nonvolatile means that by the very design of the memory, a means exists by which the contents of the memory remain intact without an external power supply.

The segments of memory in a programmable controller system are straightforward.

Application Memory. Also called logic memory, it is the section of memory used to store the actual control program that the controller uses to control the manufacturing process. This control program is usually created by the system user.

Data Table Memory. This term collectively refers to the variable (register) memory, and the input/output status or image tables. The variable memory contains timer and counter values, along with any data used in mathematical calculations performed by the application program. The I/O image tables contain, as the name suggests, a representation of the actual input/output point status, either on or off.

Executive. Also called system firmware (or just firmware), this section of memory contains the base intelligence of the system. The executive program supervises the basic chores of the programmable controller system including communications with subsystems, control program interpretation and execution, CPU diagnostics, and other housekeeping tasks included in every scan.

Scratch Pad. This is a temporary memory area used by the system to store the step-by-step and interim results obtained through some calculations. In some systems, the scratch pad memory contains the programmable controller statistics, such as memory size, amount used, and any active diagnostic flags set.

Various segments of the programmable controller use different memory types to accomplish different design or application purposes. Below, we shall examine a sample of memory types, and contrast their use in programmable controllers.

Read Only Memory (ROM). This memory was one of the first commercially viable nonvolatile memory types used in microprocessor-based systems. ROM get its name from the fact that the memory can be read from (information extracted), but cannot be written into (information placed in). A number of manufacturers of programmable controllers use ROM memory to store the executive programs. This is because it normally requires no adjustment or editing once the system is shipped from the manufacturer. ROM is rarely, if ever, used as application memory, and cannot be used as data table or scratch pad memory because it cannot be updated with data from the operation of the programmable controller execution.

Random Access Memory (RAM). This is a volatile memory, but has the advantage over ROM of being capable of being written to as well as read from. It is for this reason tht is is sometimes called read/write memory. Any location within the memory can be accessible. Because it is volatile, the memory contents will be lost if power is lost. With a properly designed battery backup system, RAM can retain its current contents during extended power outages. Most RAM today is of the Complementary Metalic Oxide Semiconductor (CMOS) type, and uses very little power. It is this combination that makes RAM the memory of choice for most of today's programmable controller systems, especially for use in the application, data table, and portions of some scratch pad memories. RAM's volatile design is offset by its high speed and low cost.

Programmable Read Only Memory (PROM). PROM is a type of ROM that is programmable with commercially available equipment. Once programmed, however, it cannot be altered and hence is nonvolatile. It would therefore be generally unsuitable for use as application or data table memory. In some programmable controllers, PROM is used as application memory, but has a RAM memory used for the portions of the program that might change later. It could, however, be used as executive program storage memory.

Erasable Programmable Read Only Memory (EPROM). This is a category of PROM memory that can be erased after initial programming and reprogrammed for another use. The erasing process is performed with the use of ultraviolet light applied to the lens window in the top of the PROM integrated circuit package. EPROM is sometimes used as executive and scratch pad memories, and can be used as application memory with a complementing RAM memory to act as operational memory. Many times a machine builder will design a machine control program and set the completed, debugged results into EPROM. This provides the end user with a nonvolatile system that the machine builder is sure will not easily change. It would not be appropriate to use EPROM if multiple changes are required at the user's site after installation of the machine.

Electrically Alterable Read Only Memory (EAROM). This category of ROM memory is actually quite similar to PROM, but is electrically erasable using voltage applied to the integrated circuit instead of ultraviolet light. While a useful form of nonvolatile memory, the EAROM requires RAM as backup when used as application memory.

Electrically Erasable Programmable Read Only Memory (EEPROM). This remarkable nonvolatile memory combines the attributes of EPROM with the speed and flexibility of RAM. EEPROM can be programmed and reprogrammed with the standard programming device supplied with most programmable controllers. Because of this fact, it is being used in some of the newer programmable controllers as application memory. It could also be used as scratch pad and data table memory, and even as executive memory if precautions are taken to prevent accidental erasure. EEPROM has a slight delay involved in the reprogramming process, and has a finite reprogramming life. In spite of these minor drawbacks, EEPROM has many outstanding benefits.

Core. Core memory is a relatively mature memory technology that is nonvolatile by design. It was one of the few memories commercially available for the first programmable controllers. It works by using a tiny toroidal ferrite coil which is energized to store each bit of information. An electrical pulse is used to "set" each coil to a zero or a one. Although core was a good functional solution for its day, the speed and cost improvements of the newer semiconductor memories have attracted most of the manufacturers of programmable controllers today. There are still a few controllers today that offer core memory as an alternate to RAM. Figure 4.4 illustrates a core memory segment.

As we saw in Chapter 3, programmable controller memory is formatted into bits, bytes, and words of memory. A bit is a single storage element for either a zero or a one. A byte consists of eight bits, and a word (normally) consits of 16 bits, or two bytes. Some systems still use a word length of eight bits, but most have adopted a 16-bit word, even though they may use an 8-bit microprocessor. Figure 4.5 shows a 16-bit word format.

Depending on the specific design of the programmable controller, it will have a stated memory capacity. This is an indication, although not the only one, of the capability and power of the system. Medium and large controllers are normally expandable from one memory size to their maximum size. Small controllers are normally fixed in their memory size. Size of the memory capacity must be examined relative to the word size (8- or 16-bit) and utilization. While it is clear that twice the information can be stored in a 16-bit word than in an 8-bit word, it may not be immediately clear that some controllers utilize memory more efficiently than others. For example, a normally open contact and its associated reference address (i.e., Input 1), may use one 8-bit byte each for storage. Combined, they consume one 16-bit word. Some controllers may use more memory than this for these instructions or others. In a large program, these inefficiencies can build on each other to cause a poor utilization of the system memory. A careful analysis of the various programmable controller models is required to assess utilization efficiency. Normal practice calls for an additional 20 to 40% of memory size to be specified to allow for modifications and later expansion. This analysis, combined with knowledge of the application needs, will allow for an intelligent choice of programmable controller.

The memory of a programmable controller is organized in what is called a memory map. This segments, through a process known as partitioning, the memory into functional units. Figure 4.6 shows an example of a typical memory map organization. All manufacturers use a slightly different technique in designging their controller's memory map. Some have variable partitions while

STORE "0"   STORE "1"

Figure 4.4   Diagram of core memory.

others are fixed. All, however, are designed to segment the following functional areas:

> Executive program(s)
> Scratch pad
> Input/Output image tables
> Data tables
> Application program

We will now elaborate, in overview fashion, on each of the memory map segments. As noted earlier, some controllers offer the user the flexibility (sometimes considered a constraint), of being able to vary the partitions within the memory map. This, in

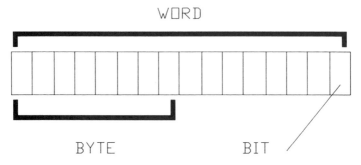

Figure 4.5   Memory word format diagram.

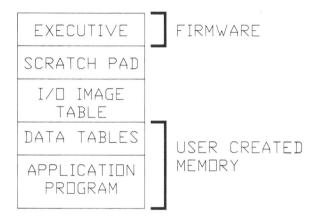

Figure 4.6 Memory organization example.

essence, allows the user or system builder to customize the sizes of the application, data table, and other memory segments to suit the particular application. Other controllers offer a preconfigured system, making assumptions about appropriate sizes for the various memory segments and their associated partititons. This eliminates the need for the user to deal with this sometimes confusing operation. As the architecture of programmable controllers continues to evolve, it is likely that the variable partition method will gain favorable momentum. This is likely because it accommodates a wider variety of operating systems and application programs since it can be tailored more effectively. This more flexible future may include the ability to utilize a standard hardware from many manufacturers with special application software to accomplish an industry specific solution.

Executive. This is the basic intelligence of the programmable controller. It allows any application program instructions to be interpreted and acted upon. It is transparent to the user and is almost never considered to be included in the manufacturers rated memory sizes.

Scratch Pad. Also transparent to the user, this memory allows interim computations and some system configuration parameters to be established.

Input/Output Image Tables. This is one of the most basic and straightforward segments in the memory map. This section of memory contains a stored representation of both the internal and external I/O "points." An internal point is an input or output that is used only in an internal control logic process, and is not directly associated with the physical I/O modules. An external point is one that is directly associated or mapped to a physical I/O module, which in turn is physically connected to a sensor or actuator. These tables of the I/O are accessible and viewed by the programming device and some other programmable controller peripherals. They can then be observed or manipulated directly for program creation, editing, or later troubleshooting after the system has been installed. This memory segment is normally partitioned to some default value corresponding to the maximum I/O capacity of the programmable controller. The view seen on the programming device screen is the most current information on the status of the I/O, as it changes per the application program instructions and real world environment.

Data Tables. Sometimes called the register tables, this segment of system memory contains the variable references used in the execution of the application program. Formatted in 16-bit words (8-bit in older systems), this would include storage of timer and counter accumulated values, in some cases timer and counter preset values, variable storage references for mathematical functions, storage of analog values converted to digital, storage of BCD or ASCII information, and so on. This segment, in controllers that allow it, is sized by the user to trade-off application memory size for register table capacity. This becomes a perplexing issue in systems that are particular data intensive at the expense of application program size.

Application Program. This segment contains the actual ladder logic control program. Hence it is sometimes called the logic memory section. Again variable in size for some systems, it is created, edited, and later viewed during operation with the help of the programming device. A section of ladder logic is created with the programming device using contacts, coils, and other references, and then is converted to machine level code for use by the central processing unit. There are many techniques and devices to accomplish this task, and we will examine some of these in Chapter 6.

4.6 POWER SUPPLY

As smart as the central processing unit is, it would be nothing without good, clean, reliable power. The manufacturer of the programmable controller system takes special design and manufacturing care with the system power supply. Some designs allow the power supply to be used external to the main CPU chassis, while others make it an integral part of the system chassis. Figure 4.7 shows a typical system power supply. In spite of these physical differences, the primary function remains the same: that is, to provide a consistent level of clean, low voltage direct current (DC) power to the system electronics, and protect the system from normal line voltage fluctuations.

Most of the programmable controllers used today employ alternating current (AC) with voltages of 115 or 230 as line voltages. In some unique applications, however, 24 volt or 120 volt DC are used. This is especially true in systems that utilize batteries to back-up an unreliable utility power system, or to suffice where normal utility lines do not exist. Examples of this include standby power generators used in remote areas to provide electrical power, and railroad trackside control applications where system availability and integrity is critical. Another interesting example is that of generating facilities that use 120 volts DC to insure full-time control of the generation process inspite of power changes and fluctuations.

The power supplies on most programmable controller systems can operating and provide consistent power inspite of normal power line fluctuations. Typical supplies will provide "business as usual" for fluctuations in voltage of 95 to 130 volts AC for a 115 volt nominal system, 190 to 260 volts AC for a 230 volt system, and can accommodate frequencies from 47 to 63 hertz. In some extreme examples uninteruptable power supplies (UPS) or constant voltage transformers are used to guarantee constant power to the control system.

In a different environment, isolation transformers may be appropriate. Some installations may have fairly reliable power, but experience high incidents of electromagnetic interference (EMI) generated from adjacent equipment. Good judgment in the location of equipment and a control system is warranted of course, along with proper control cabinet choice. Where this nor-

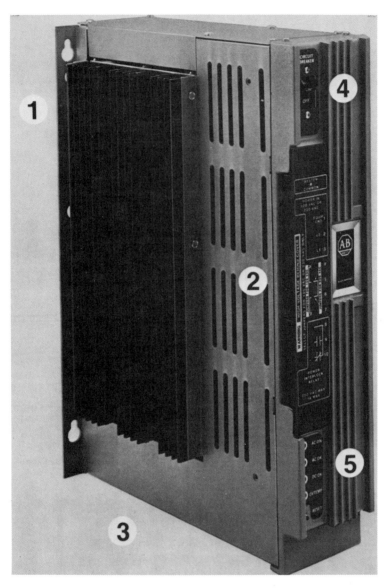

Figure 4.7 Photo of power supply for programmable controller. 1, rugged casing; 2, terminal connections under protective cover; 3, processor I/O connections located underneath; 4, circuit breaker switch; 5, diagnostic indicators and overtemperature reset. (Courtesy of Allen Bradley.)

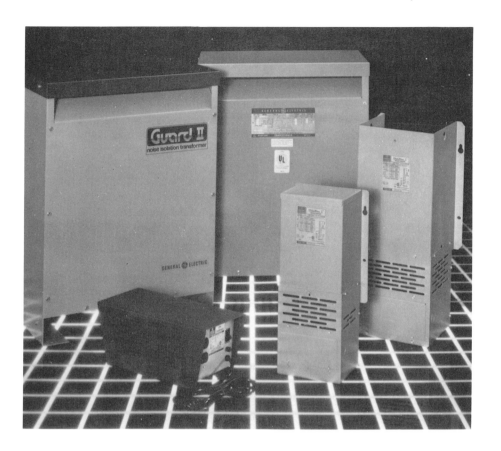

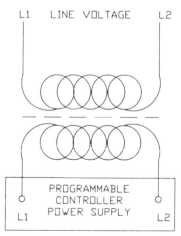

Figure 4.8 Isolation transformer diagram and photo. (Photo courtesy of General Electric.)

mal course does not suffice, however, an isolation transformer
may prove to be the best solution. Connected between the programmable controller and the power source, the isolation transformer keeps most stray noise off the line to the controller. Figure 4.8 shows how a typical isolation transformer might be connected and used.

4.7 DIAGNOSTICS

Central processing unit diagnostics are designed in most systems
to allow the unit to detect and report a number of ills, some internal to itself, and some external. In some systems, these detected
problems are used to close a hardwired power supply relay for
alarm purposes, in other cases internal coils are set. Included in
the list are:

> Total scan time exceeds watchdog timer value
> Unrecoverable communication error to subsystem
> Memory backup battery low or failed
> Parity error
> CPU failed executive self-check
> Input line voltage out of acceptable range

Depending on system size and construction, errors are announced on a board basis by front mounted light emitting diodes
(LEDs). This allows rapid pinpointing of the likely problem, and
quick replacement of the suspect board or power supply. The
faulty board can then be repaired or returned to the manufacturer for replacement. In some extreme cases, emergency installation of a complete standby CPU with resident program may be
more expedient than troubleshooting and board-swapping a CPU
on line.

5

The Input/Output System

In spite of the incredible gains made in programmable controller systems over the last few years, we must not forget where the story began...with the Input/Output system.

5.1 I/O ADDRESSING

The input/output system is the arms and legs of the programmable controller system. Their proper operation is absolutely critical to the successful operation of the system. Without the need to interface with and control real-world devices, there would not have been a need for the programmable controller. Relay-based control systems were, by-and-large, predictable in their interactions with sensors and actuators in their daily life on the factory floor. While relay coils were difficult to install and later troubleshoot, they were (and are), incredibly rugged in their ability to withstand electrical abuse. Relay contacts were also generally well suited to the harsh factory environment. The major drawback, as we noted before, was their inability to adapt to change, as a factory's need for control processes changed.

Enter the programmable controller. With its relatively delicate electronics, it needed a reliable way to connect all of the sensors and actuators to the system. It is the I/O system that sepa-

rates a programmable controller system from an industrially
hardened general purpose computer. Today's programmable controller I/O system is designed to provide maximum protection of
the CPU and its subsystems while delivering efficiently the information drawn from and brought to the system's real-world devices. Optical isolation from odd voltage spikes and contact
"bounce" filters are routine design tactics to deal with this challenging problem. This chapter will examine the Input/Ouput
system in depth, including its installation, addressing, and diagnostics, as well as day to day use of standard and special purpose
I/O.

Programmable controller I/O is divided into two broad
classes: those with fixed or nonflexible addressing schemes and
those with flexible, adaptable addressing schemes. Addressing is
the way the control program in the CPU relates to a particular
real-world sensor or actuator. It is in this way that a pushbutton
or starter relates in the CPU memory. The design of the programmable controller dictates whether or not a system is capable of
being addressed flexibly, or is rigid in its addressing method.
There are advantages and disadvantages to both approaches. If
an installation is to be distributed in nature, or will be built and
installed in several sections or stages, a flexible addressing scheme
is generally best. This allows the system designer(s) to create the
control logic software without being constrained to follow a sequential I/O assignment, resulting in a randomly addressed and
installed I/O system. Some designers, however, object to this
flexibility. If a system is very large, or there are many different
systems in a manufacturing facility, such a design may make it
difficult to diagnose and correct problems when they inevitably
occur. This is especially true when good documentation of the
programmable controller is not provided.

Addressing of the I/O is done in two primary ways, decimal
and octal. The decimal method is the straight sequential assignment of I/O points, such as Inputs 1 through 8 for the first input
module, Inputs 9 through 16 for the second, Inputs 17 through 24
for the third, and so on. Octal, as you recall from Chapter 3, is a
numbering system based on 8, so an octal addressing scheme
would use the numbers 0 through 7. In our 8-point input module
example above then, the first module would have addresses 0
through 7, the second would have addresses 10 through 17, and

the third would have addresses 20 through 27. In flexible systems, I/O is addressed by either physically setting a series of rocker switches associated with a certain slot in the chassis, or some newer systems use EEPROM memory to contain I/O address data. The addressing is normally established during the initial system configuration, and then is permanent for the life of the system. In nonflexible systems, the individual slot and point addresses are normally dictated by the sequence in which the I/O chassis are connected together, or in the case of small programmable controllers, contains one chassis and hence has I/O addressing fixed by design. Figure 5.1 shows a fixed I/O programmable controller. Figure 5.2 shows a typical I/O addressing configuration, along with how the input and output points relate to the CPU memory image tables.

Figure 5.1 Photos of fixed I/O chain programmable controllers. (Courtesy of General Electric.)

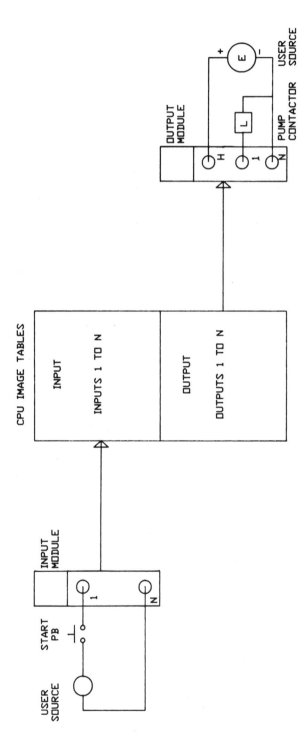

Figure 5.2 Diagram of typical I/O addressing schemes.

5.2 EXTERNAL WIRING—CONNECTIONS TO SENSORS, ACTUATORS, AND POWER

Recent studies performed by a major electrical and electronics manufacturer concluded that many of the difficulties experienced with using programmable control systems came from the external wiring to the sensors, actuators, and power applied to the I/O modules. This result came in two parts: (1) that of the initial installation wiring with erroneous operation or disastrous power-up results from improper wiring, (2) or after operation begins and troubleshooting for an intermittent fault begins. It is for that reason that we will examine some of the more common techniques involved with I/O wiring. Connections to sensors will be treated first, and then connections to actuator. Power connections will be integrated into the text of each as appropriate.

The most common sensors used with programmable controller systems are pushbuttons, limit switches, proximity switches, and in the analog world, thermocouples and RTDs. Figure 5.3 illustrates the wiring methods used to wire discrete sensors, such as pushbuttons, to a typical programmable controller. Figure 5.3(a) is a grouped input module example. That is, a common is provided for a group of input points, usually four, giving two groups of four on an 8-point module. This is a cost effective way to have sensor connections where the input voltage source is common and does not require isolation from one point to another. Another way to decrease the cost per I/O point used is to use the higher density 32-point modules now available from some manufacturers. This allows most sensors to be connected to the system in a manner that brings 2:1 returns or better. Where voltage source isolation is required, an isolated input module is warranted. Figure 5.3(b) shows typical wiring for such a module.

Analog input modules are wired and used quite differently. While there are similarities in shielded cable installation practice and calibration procedures, each manufacturer of analog input modules uses a slightly different technique for wiring depending on module design and objectives. Figure 5.4 shows a typical analog input module wiring scheme. Included is the different approaches for current (i.e., 4 to 20 mA), and voltage (–10 V to +10 V). A variation on this method is used with low level analog inputs such as those provided by thermocouples and RTDs. Both

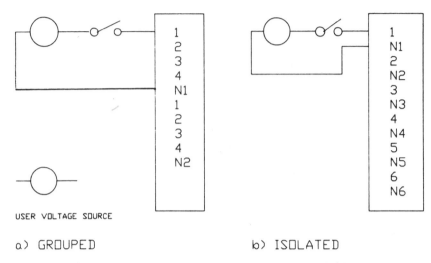

a) GROUPED   b) ISOLATED

Figure 5.3 Diagram of discrete input module wiring: (a) grouped; (b) isolated.

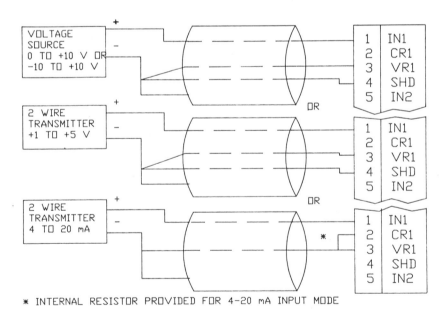

* INTERNAL RESISTOR PROVIDED FOR 4–20 mA INPUT MODE

Figure 5.4 Diagram of analog input module wiring.

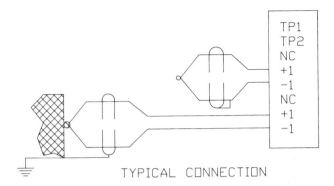

TYPICAL CONNECTION

Figure 5.5 Diagram of thermocouple module wiring.

are designed to sense temperature and are used in different applications depending on ambient temperatures and other factors. A thermocouple input module with typical wiring is illustrated in Figure 5.5. Special I/O modules, which will be examined later in this chapter, call for unique wiring methods. Consider as an example the Motion Control Module shown in Figure 5.6.

Actuators have their own set of rules and practices with regard to wiring. Discrete output devices, such as contactors, starters, solenoids, and pilot lights, are served by an output version of the discrete input modules described above. Figures 5.7(a) and 5.7(b) show typical wiring for those cases where grouped commons are allowed, and those where individual commons are required. Relay output modules are used where current loads are higher than that supported by the standard output module, or where a mix of voltage and current loads must be handled by one controller system. Figure 5.8 shows a relay module wiring diagram. Relay modules are also useful to create multiplexed inputs from a variety of sensors to a single analog input module, for example. As with the input modules, 32-point output modules are available to reduce the per point cost. Figure 5.9 shows typical wiring for an analog output module, complementing the input example that we saw earlier.

One of the most exciting current developments in input/output technology is just emerging on the scene—a mixed input/output module. It has two primary attributes: (1) the ability to be distributed in an I/O system all the way down to the sensors and

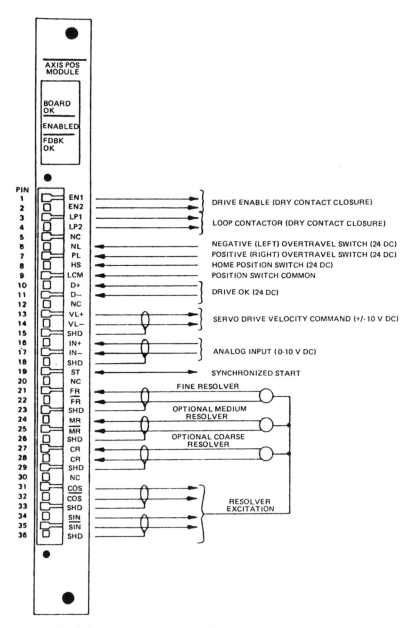

Figure 5.6 Diagram of motion module wiring. (Courtesy of General Elec tric.)

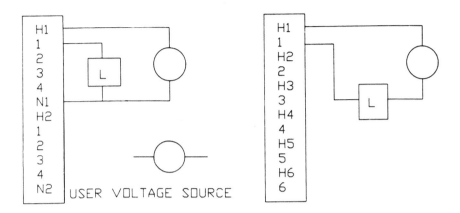

Figure 5.7 Diagram of discrete output module wiring: (a) grouped; (b) isolated).

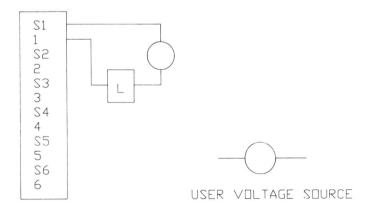

Figure 5.8 Diagram of relay module wiring.

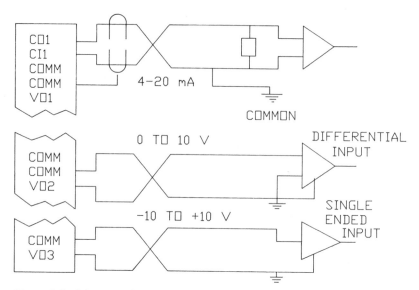

Figure 5.9 Diagram of analog output module wiring.

actuators, thereby reducing the installation costs of a control system, and (2) the ability of the I/O to detect failures, both of itself and the sensors and actuators connected to it. An attractive by-product of this new technology is the ability to have a mixture of inputs and outputs on one module, with each point on the module selectable as either. This allows an 8-point module to be configured as seven in and one out, four in and four out, or two in and six out, for example. A wiring diagram of this new I/O is shown in Figure 5.10, and an example is shown in Figure 5.11. Some of the programmable controller manufacturers offer this advantage now in the form of a dedicated four-in-four-out module, but without the added benefits of cost-effective distribution and enhanced diagnostics. More on this breakthrough later in this chapter.

5.3 INPUT/OUTPUT MODULES

One of the critical choices made when configuring a programmable controller system is that of the I/O module choices themselves.

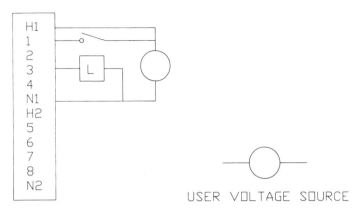

USER VOLTAGE SOURCE

Figure 5.10 Diagram of mixed input and output module wiring.

Figure 5.11 Photo of I/O with individually addressable points. (Courtesy of General Electric.)

While each manufacturer has many of the basics covered, there are a great many details to be attended to when selecting the various modules for your system. This is becoming even more important with the newer, more data intensive modules available. This includes the ASCII, Basic, and Motion Control modules, as well as others. These microprocessor-based modules have their own set of instructions and eccentricities, along with an inch-thick manual for programming and operation. Even the more simple modules offered come with basic needs for installation care, including some configuration in the form of jumper setting on some models.

For this reason, Sections 5.3 through 5.5 will provide a survey of the general types of I/O modules available on the market today, along with some of the advantages and compromises required with certain designs. Section 5.3 will begin with the discrete I/O, followed by analog I/O, and special I/O. Remote I/O is covered in Section 5.4, and diagnostics (a premier issue in today's systems) in Section 5.5.

5.3.1 Discrete I/O Modules

While all discrete I/O modules are not created equal, many of them do share the same general functions and features. The general requirements of a discrete I/O module is to take logic level signals from the I/O communication bus and control and convert them to field level signals, or voltages, at the actuators, with the reverse process occurring at the sensors. Most of today's I/O modules reside in the slot or slots of a rack or chassis that contains several slots, and a power supply either internal to the rack or external to it. Power from the supply is used to power the electronics, both active and passive, that reside on the I/O module circuit board. The relatively higher currents required by the loads of an output module are suppled by user-supplied power. A typical discrete I/O module is shown in Figure 5.12. Over the years, a number of features have come to be considered important and, in some cases, required as standard equipment by programmable controller users and system integrators. These include:

> *Input Modules.* LED or other visual indicators to indicate point status; filtering circuits on board to eliminate "contact bounce"; and voltage isolation (usually optical) to pro-

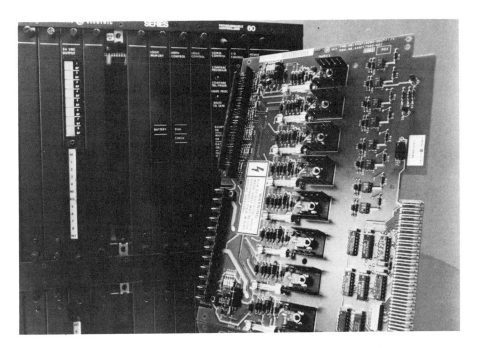

Figure 5.12 Photo of a typical discrete I/O module. (Courtesy of General Electric.)

vide protection of the electronics from the harsh electrical world outside. Heavy duty connectors are provided to allow connections to both field connections and backplane.
Output Modules.   These also have indicator lights to indicate point status, although these may be neon type if used on the field side of the connections. Fuses are generally required here, and are provided on a per circuit basis. This allows each circuit to be protected and operated separately. Some modules also provide visual indicators for fuse status, a real time-saving feature. Newer modules provide circuit protection on a programmable nondestructive basis, eliminating the need for fuses. Voltage isolation and transient protection are also provided. The same heavy duty connections are provided here also.

Discrete I/O is available from a variety of manufacturers in
a number of configurations. The first choice is the voltage required. The range to choose from includes AC voltages from 115
to 230 volts, and DC voltages from 5 volt TTL levels, 12 volt,
24 volt, 48 volt, and even 120 volts from some manufacturers. In
the DC versions, a sink or source current option is often offered.
This has to do with the current flow direction convention, and the
nature of some sensors and actuators. The next area of choice is
the number of points on a module. This can range from as few as
one to as many as 32. The average number seems to be eight,
especially on AC modules. The higher density modules can have a
dramatic impact on rack or panel space, and therefore system cost.
As a final consideration, discrete AC modules are available from
some manufacturers in an isolated version, allowing use in motor
control centers and other environments where individual supply
connections are required for each point.

Many of the discrete I/O modules offered on the market today have common user-driven features. The mechanical arrangements of the I/O module circuit board and faceplate require that
the faceplate to which the field wiring is attached be removable
from the circuit board. This is designed to make servicing the programmable control system easier, since disturbing the field wiring
on an installed system is inviting trouble in the form of errors in
rewiring with sometimes disastrous results. Most faceplates and
I/O modules are keyed to prevent putting the wrong faceplate on
the wrong module. This can keep electrical surprises to a minimum. Wiring connections on the faceplate are normally of the
box lug terminal type, and allow wire up to No. 12 AWG to be
attached easily. UL Standard 230 C is normally used as a guide-
line for wiring and connection practices. Normal faceplates include space designated for labeling the physical I/O points for
their real-world significance. This is adjacent to the status indicators for easy troubleshooting.

There are, in addition, a number of electrical specifications
common to many discrete I/O modules. Included here would be
an On and Off voltage range associated with the two logical states.
Typical for a 115 volt Input module is 90 to 130 volts as an On
range, and 0 to 30 volts as an Off range. Also important in the
electrical specifications are the On and Off delay times. This is a
function of, among other things on the input module, the filtering

electronics values, with typical values of 10 to 20 milliseconds On delay, and 20 to 50 milliseconds Off delay. Because the output modules have no need for filtering, the On delays are much shorter, as low as 1 millisecond. Off delays are a function of the board electronics, with Triac driven outputs being limited to a one-half frequency cycle. Input loading impedance and output leakage current are considerations in some systems and the reader is left to individual system and manufacturer analysis for this matter. The vast majority of programmable controller-based systems will perform perfectly without consideration to these details. However, some high-speed applications, or those with unusual sensor and actuator requirements, will need special attention to these electrical criteria. Some manufacturers provide jumper selectable input filtering options, allowing customization of sensor and input integration. This is especially common on DC and TTL input modules.

5.3.2 Analog I/O Modules

Analog input and output modules are utilized whenever the sensors and actuators chosen are of the continuously variable type in contrast to the discrete I/O on-off type. In this class of sensor and actuator would be flow measuring instruments, specialized position sensing devices, strip chart recorders, and variable speed drive systems. Also included is a whole host of sensors and actuators that have had signal conditioning performed to bring their levels to the so-called "high-level" signals. Figure 5.13 shows a typical analog I/O module. Most of today's analog I/O is provided in three ranges: (1) 0 to 10 volts (DC), also called unipolar; (2) –10 to +10 volts (DC), also called bipolar; and (3) 4 to 20 milliamp (mA DC), also referred to as current. This last version can many times also be expressed as 1 to 5 volts (DC) with a simple reconfiguration of the module. Like their discrete counterparts, analog I/O modules require power to be supplied from the I/O chassis in which they are mounted. Input modules will normally have eight channels available, but designs are available that provide four or even two channels. Output modules are available in configurations of four or two channels. Newer breakthroughs in analog I/O provide the mixing of analog inputs and outputs on one module, normally providing four input channels and two output channels.

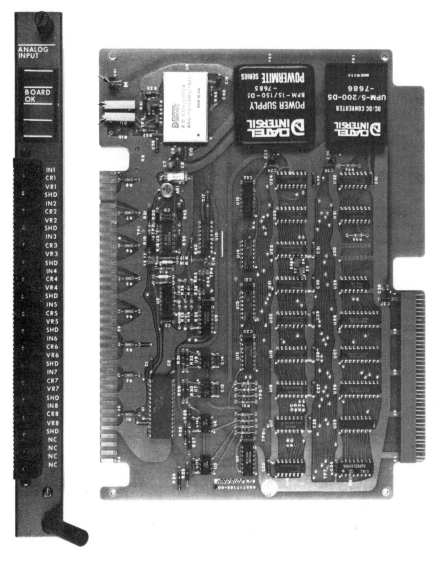

Figure 5.13 Photo of a typical analog I/O module. (Courtesy of General Electric.)

Analog modules are generally classed by their resolution, which is their ability to convert the analog information to digital form in the case of an input, and the reverse for an output. This resolution is expressed in bits; for example 12-bit or 8-bit resolution. This would correspond to 1 part in 4096 and 1 part in 256, respectively. It should be clear that the higher the resolution, the more accurately an analog value can be represented digitally. If you think of the analog signal as being physically continuous, representing exactly the physical measurement of action occurring, it is easy to visualize the representation of that in digital form. Discrete steps must be used to illustrate the continuous nature of the analog signal. Therefore, the higher the resoltuion, the more steps are available, and the more accurately the signal can be converted. And while accuracy and resolution are certainly related, accuracy of a module is actually how well and intact a signal passes into and out of the programmable control system, and is a function of the entire board design. It is normally a function of temperature, and is expressed as a percentage of full scale at a given temperature, for example, $< +/-0.025\%$ of full scale at 25°C. Other considerations when evaluating analog I/O are:

Power Requirements That is, does the module draw a large amount of the available power from the power supply, or require special external power supplies.

Input Impedance and Capacitance. These are used to match the sensor to the module and are expressed in mega ohms and pico farads.

Common Mode Rejection and Cross Talk. Expressed in dB at a certain frequency, these refer to the module's ability to prevent noise from interfering with the data integrity, both on a single channel, and from channel to channel on the module.

Temperature Coefficient and Total Output Drift. These refer to the module's ability to remain accurate over a certain temperature range, and are normally expressed in the form of fractions of full scale per degree of change. An example of this would be <6 PPM per degree centigrade.

Mechanical considerations for analog modules are similar to discrete modules, with only a few differences. Most manufacturers only provide an indicator on the faceplate to determine the circuit board's health, while only a few provide any indication of the present value of a given analog channel. Those few that do are quite innovative, and provide a visual indication of the current input or output value of the module. This is accomplished through a series of discrete LEDs that provide the binary representation of the analog value. As with the discrete modules, space is provided for the user to attach labels representing the real-world significance of each channel.

Electrically, the analog modules are quite different from discrete. Most have a single Analog/Digital or Digital/Analog convertor on the circuit board. This requires that the channels share that convertor through multiplexing. This is normally accomplished with a combination of board design and CPU ladder logic. Open wire, overrange, and underrange conditions are detectable, and are made available to the CPU as diagnostics data. The newest technologies provide for a digitally designed analog module that eliminates calibration problems and provide automatic linearization and engineering unit scaling at the module level. This means that when data is received by the CPU it has been converted to values meaningful to the physical parameter being considered, that is, flow, position, etc.

5.3.3 Special I/O Modules

The architectures and evolutions of programmable controllers have been almost totally concerned with the task of I/O control through the solution of ladder logic. It is not too difficult to understand that the designs have, therefore, been generally optimized around this task. With the broadening scope of programmable controller applications in today's industrial control arena, these "bit" manipulating I/O directed designs are not always suitable in their new roles. The deficiency shows up in the controller's inability to efficiently handle large amounts of data of different types. For example, many of today's sophisticated controllers have the ability to handle integer data in blocks of 128 or 256 words fairly efficiently with the ladder logic instruction set provided by the manufacturer. This same system, at least presently, will probably not be able to tackle floating point data, that is, data with one or more significant decimal places, or data blocks several hundred words in length. Because of this deficiency, many of the programmable controller manufacturers have developed special purpose I/O boards to address the application needs of their controllers in today's marketplace. Most of these special I/O are of microprocessor-based design; however, a few are not. We will examine both below, and will consider them in their likely application category. Categories included will be motion control, high speed counting or position sensing control, process control (PID), data management and control, and communications control.

Motion Control

There is today a category of application that involves the precise control of motion, integrated with the programmable controller system. At this point it is valuable to contrast Computer Numerical Controls (CNC) to Programmable Controllers (PC) as they relate to motion control. A CNC is a highly developed, optimized control that is designed largely for specific types of motion control for machine tools. And while they are expanding into cell management roles in some applications, their role is fairly specialized. A programmable controller is normally more appropriate for a wider range of motion control applications. These can be linear or rotary motion, or both, and can have requirements for precision of up to 0.001 in. of error relative to commanded position. Examples of machines that would use such a control option are grinders and other special purpose metal cutting, metal forming, and assembly machines. Normally these machines require a programmable controller to provide the required sequencing, while integrating single or multiple axis motion control for dedicated or at least fairly standardized machine processes.

Most of the motion control options offered by manufacturers today have similarities in design and function, but only a select few have considered the system integration and end use needs comprehensively. The basic requirements of the application are to provide an intelligent closed-loop motion control ability to effectively couple the programmable control sequencing operation with precisely executed positioning commands to a servo drive positioning system. An exception to this is the use of a stepper motor control system, which may be run in closed or open loop mode. A typical application might include three to five axes of motion control in a single programmable control system. The function normally requires one to three circuit boards per axis residing in the controller's I/O system. This occasionally is the source of a disproportionate power usage or data bottleneck in some systems, so care is required in configuring the system. Each axis's circuit board(s) contain a primary microprocessor for executing the selected motion profile, and memory for storing the motion profile or profiles on the board. The reason for this is so the motion control can go on independently from the main processor scan, and can therefore do it more efficiently. A typical motion profile is shown in Figure 5.14, and illustrates the multivelocity positioning requirements of some applications. This might be typical for some metal cutting operations where a high rate of speed (V1) is used over a portion of the machine travel, and a second, lower speed (V2) is utilized in the actual cutting process. A motion control I/O module takes selected commands from the CPU and then executes those commands, reporting back to the CPU when requested. A typical position loop may be solved and updated as often as every 2.2 milliseconds. The module takes in a current position reference directly from a position sensor such as a resolver or optical encoder. This current position information is compared to the commanded position at that time and a resulting voltage ranging from −10 to +10 volts is issued to the connected

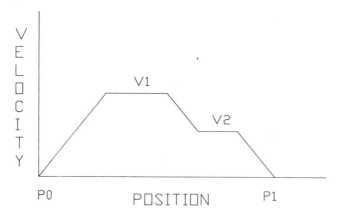

Figure 5.14 Motion profile diagram.

servo-drive system, or in some cases a stepper motor drive system. Acceleration, decceleration, and overtravel, among others, are parameters normally controlled or accepted by the module directly. Figure 5.15 shows a single board motion control module, with faceplate connections for position sensor and external drive system. Some of the more comprehensive modules include extensive diagnostic data to be returned to the CPU about the condition of the module, the drive attached to it, and any illegal conditions that might occur. Some manufacturers attack the need for motion control with a separate system attached to the programmable control with I/O lines. Such a configuration option is shown in Figure 5.16.

High Speed Counting or Position Sensing

One of the requirements in many applications in both discrete and batch process industries is the ability to rapidly count things, either pulses from the rotation of a spinning disc, or the passing of units on a conveyor or packaging machine. In some cases, this result is used to track the rotary or linear position of an apparatus. This need has driven the development in recent years of an intelligent I/O module generally called a high speed counter. This microprocessor-based module is designed to accommodate count pulses that occur too rapidly for the CPU scan to accurately pick them up. The high speed counter counts and accumulates the pulses, and makes the results available to the programmable controller CPU during the normal I/O scan window. This allows the CPU to concentrate on solving ladder logic, unhindered by the processing needs of the high speed counter. Most of today's designs allow for the input to the module of quadrature incremental encoders, digital tachometers, and mechanical or transistor switches. Incoming pulse rates of up to 50,000 hertz can be accommodated in some designs. Figure 5.17 shows a photo of a typical high speed counter module, while Figure 5.18 illustrates a typical timing diagram using a quadrature encoder with a high speed counter. Figure 5.19 shows a typical incremental encoder. Most modules have one or more real-time outputs that are energized when the accumulated count reaches or passes a predetermined preset value, set up in the software. This mechanism allows the module to handle time-sensitive applications without relying on the CPU to detect

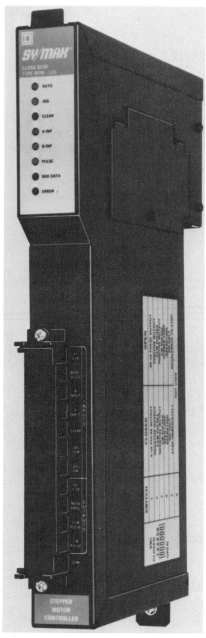

(a)

Figure 5.15 Photos of motion control modules. (a) Courtesy of Square D; (b) courtesy of Honeywell-ISSC.

(b)

Figure 5.15 (Continued)

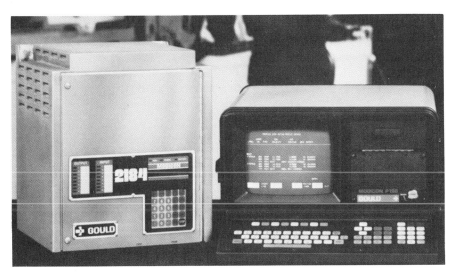

Figure 5.16 Photo of separate motion control system. (Courtesy of Gould, Inc.)

Figure 5.17 Photo of a high speed counter module. (Courtesy of Square D.)

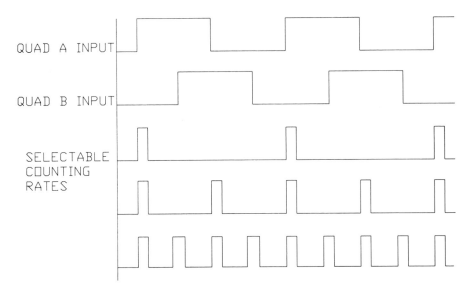

QUAD A INPUT

QUAD B INPUT

SELECTABLE
COUNTING
RATES

Figure 5.18 Quadrature pulse counting timing diagram.

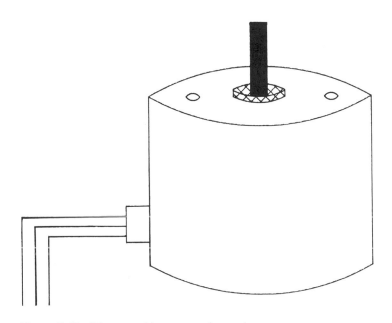

Figure 5.19 Diagram of incremental encoder.

85

the accumulated count, compare it to a preset stored in register memory, and execute another real-world output on a separate module. Applications for such a special I/O module include hydraulically driven machinery and elevator position control.

Process Control

In the early days of programmable controller applications, process control (PID) was generally performed by individual process loop controllers as stand-alone devices, or as part of a process computer. It was rarely if every done in the programmable controller. Even when analog I/O became fairly common, it was cumbersome if not impossible to do accurate process control. Process control is concerned with the closed loop control of continuously variable physical parameters. It involves the setting and regulating of these parameters, along with an alarm function when the process output exceeds a predetermined high or low limit. Process control is often referred to as PID or three-mode control, with PID (*P*roportional, *I*ntegral, *D*erivative) with each representing a tuning constant term in the equation:

$$E = SP - PV \qquad OUTPUT = KpE + KiEdt + Kd\,\frac{dE}{dt} + BIAS$$

where

$$
\begin{aligned}
E &= \text{error} \\
SP &= \text{setpoint} \\
PV &= \text{process variable} \\
Kp &= \text{proportional gain constant} \\
Ki &= \text{integral time constant} \\
Kd &= \text{derivative time constant} \\
BIAS &= \text{fixed offset}
\end{aligned}
$$

This equation is the primary equation used to perform closed loop process control. This is true whether it is performed in CPU logic, a stand-alone loop process controller, or a special I/O board designed for that function. Figure 5.20 shows a closed loop control block diagram for the PID function. In Chapter 8, we will examine stand-alone loop process controllers, and their relationship and integration with programmable controllers, but this section will deal with the PID function as a special I/O module.

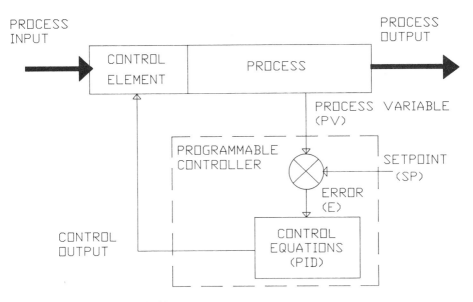

Figure 5.20 PID block diagram.

There are broad similarities and some differences in the I/O modules for PID control currently manufactured. Most of these solutions are resident in the I/O rack or chassis, deriving power and communications from the CPU from the chassis backplane. Occasionally, an external supplemental power supply is required to provide high current or unusual voltage needs. In some rare cases, an external cable is required for CPU communications. Most of PID I/O modules have at least four loops of control per module, some more. All have the ability to connect to an external control panel, to allow manual control of the loop or loops in some situations. Status and diagnostic indicators are provided to allow easy maintenance of the system once it is installed. System hardware is configured in one of two ways. One has the analog inputs and outputs providing feedback and output to the control process integrated into the PID Module, and the other requires separate input/output modules mounted in other I/O slots in the chassis. In the latter case, the current solution of the PID equation is held in CPU registers, to be shuttled in and out to the analog I/O modules as available. Configuration of the loops is required as previously noted. This includes the constants; Kp, Ki, and Kd; alarm and PV deadband, and a variety of high- and low-level alarm selections. Figure 5.21 shows typical PID modules.

Data Management and Control

In today's programmable controller applications, the needs for data management and control have never been greater. This is due in large part to the great number of special I/O modules that have been developed to accomplish this function. These are used to solve many factory floor data application needs, including interface to peripheral operator interface devices and intelligent devices such as bar code readers and other similar devices. This allows for more intelligent operation and may be useful in more unmanned situations, since the I/O module can act as a data collector and concentrator. In this way it makes some decisions based on data received, and shuttles results to other parts of the system.

Most configurations of the data I/O modules produced today are microprocessor-based boards, occupying either a single slot or multiple slots in the programmable controller chassis. They operate in one of two primary ways. One is to have only the means to convert binary to ASCII on the board along with serial port control, relying on the CPU to perform the binary register storage and data shuttling functions to the I/O module. The second is to have the functionality described above combined with the ability to perform BASIC programming and user program memory storage on the module. The second alternative is the preferred method in most applications today, as it allows any number of powerful programs to be created, controlling everything from a complex real-time operator interface CRT station to a computation-intensive batch process. It is clear to see why these intelligent options have been referred to as "a personal computer on a board." It is interesting to note the potential for special application solutions using these computer on a board options. As time goes on, it is likely that manufacturers or value-added firms will offer special data handling solutions for application or industry niches. This will broaden significantly the power of the programmable controller.

From a hardware standpoint, these I/O options are very similar. Most have diagnostic indicators to provide status and system troubleshooting aid. On-board memory on some of the more

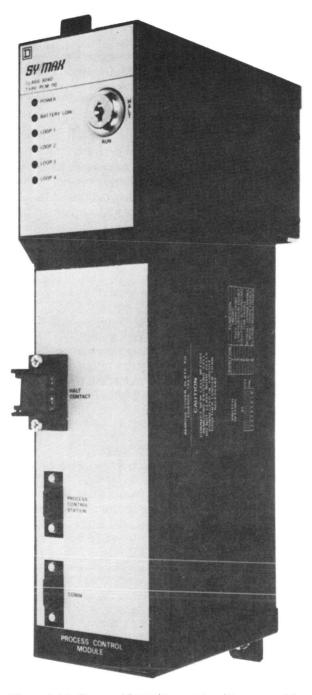

Figure 5.21 Photo of PID I/O module. (Courtesy of Square D.)

sophisticated versions can range as high as 40,000 bytes, or 20,000 words, and provide serial ports, ranging from one to as many as four. These ports can normally be configured as either RS232C or RS422, conforming to EIA standards. Software for these modules is normally provided on the board as firmware, and includes the ability to handle serial port configurations as well as the creation, storage, and execution of BASIC programs to interact with external devices. The software also provides for communication through the I/O chain, and thereby provides access to all of the system I/O and registers. This is perhaps the most powerful and differentiating feature of these modules, as this allows complete, efficient integration of the data management and control with the balance of the programmable controller system. Figure 5.22 illustrate several examples of data management and control I/O modules.

Communication Modules

In almost any serious discussion of programmable controller system applications the subject of communication comes up. In Chapter 12 we will examine communication networks, but here we will briefly cover the windows to the communication networks, the communication modules. These are generally a special category of module, and not always referred to as I/O modules since they may occupy a dedicated slot in a system chassis. Similar to the data management and control functions listed in the previous section, the communications module contains a microprocessor to accomplish communications interfaces with other intelligent devices, such as other programmable controllers or host computers. This is accomplished by installing in the firmware a communication protocol. This protocol directs the manner in which the system speaks to the outside world, or is spoken to from the outside world. It may also in some cases be configured to communicate in a method known as unformatted ASCII. This is sometimes used to accommodate odd intelligent devices, but this task is generally approached with a data control module. The software on the module allows a communication session to be arranged, executed, and confirmed between two devices. Data from the programmable controller's register, I/O, or logic memory tables can be transmitted and received using the communica-

Figure 5.22 Photo of ASCII basic module. (Courtesy of Square D.)

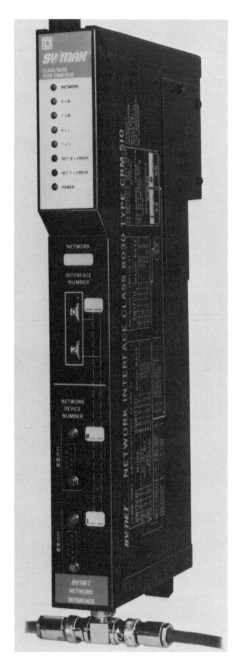

Figure 5.23 Photo of communication module. (Courtesy of Square D.)

tion module. Figure 5.23 shows a typical communications module. Note that diagnostic indicators are provided to allow easy monitoring of the module status.

5.4 LOCAL I/O VERSUS REMOTE I/O

Depending on the application and configuration requirements of the particular system being used, a choice is available regarding the I/O channels of the programmable controller. Below I briefly outline the two basic alternatives, and later in the section, will examine how the two systems would go together, and the advantages and disadvantages of each.

Local I/O. This is generally where the application calls for the I/O and CPU to be mounted together in one control cabinet. An example of this would be a single piece of relatively compact machinery, such as a grinder. Many manufacturers provide for this type of application with I/O control systems and the supporting cables for *parallel* I/O communication. These cables may have as many as 32 conductors. This is merely an extension of the system bus communication, and has many advantages.

Remote I/O. Also referred to as *serial* I/O, this alternative has the ability to be used where applications call for the distribution of the I/O chassis' over large distances. A typical example would be an automatic storage and retrieval material handling system with many hundreds of sensors and actuators scattered about. The cabling is much simpler here, a four conductor arrangement normally, but most systems require special remote I/O driver and receiver boards to accommodate this cable.

Depending on the model and size of a programmable controller, a manufacturer may offer one, or the other, of these I/O communication options, or may offer systems in which both co-exist (Figure 5.24 shows a diagram of such a system). The local or parallel I/O option allows rapid, in some cases synchronous, updating of the real-world I/O. Also the term local is a bit deceiving in some systems, as they may allow "local" I/O to be as far as 2,000 feet from the CPU. These systems also normally allow access from the programming device at any point along the parallel I/O chain, which can be extremely convenient during sys-

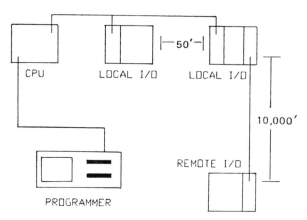

Figure 5.24 Diagram of local and remote I/O system.

tem configuration and troubleshooting. Multi-conductor cables can be quite expensive and difficult to install, especially in an existing facility. The remote or serial option offers a lower cost and a less complex cable arrangement, but requires special drivers and receivers to operate. It is likely that as systems continue to evolve, and needs change, that the distribution of I/O will continue to become more efficient. Most of today's systems require a remote I/O system to have traditional racks and power supplies. One of the newer approaches uses "blocks" or remote I/O points installed very near the sensor and actuator points, and eliminates the need for separate racks and power suppliers. Figure 5.25 shows such a system.

5.5 DIAGNOSTICS

Diagnostics for the programmable controller I/O system is one of the most important, and perhaps the least well understood issues on the factory floor today. Studies show that the I/O is the most likely subsystem to experience problems in today's controllers. This fact is complicated, however, by the fact that many of the faults or problems attributed to the programmable controller I/O system are actually problems of the sensors and actuators con-

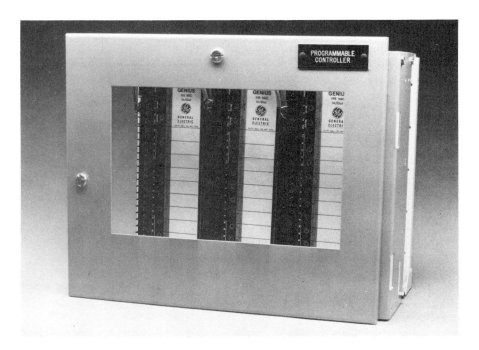

Figure 5.25   Photo of remote I/O block. (Courtesy of General Electric.)

nected to the I/O. Another contributing factor is the wiring
integrity of the sensor and actuator. Many of today's programmable controllers have automatic or semiautomatic means by
which faults or failures in the I/O system can be detected and
reported. These work in a variety of ways internally, but generally
produce the same desired results. A good diagnostics system will
include the ability to detect faults in the I/O chain from a communications parity error, from a loss of I/O communications in
the chain, a failure coming from a failed rack or rack power supply; some even detect a failed I/O module. These faults are reported in a variety of ways, from indicator lights mounted on the
front of the affected I/O module, to diagnostic messages generated
on a stand-alone display or CRT programming device.

   One new technology approaches the diagnostics issue on a
system basis. That is, not only are the programmable controller

Table 5.1 Diagnostics Statistics

Fault type/source	Fault distribution	Faults currently reported	Potentially detectable
CPU Parity error, processor, etc.	5%	100%	100%
I/O-Power supply, input/output module failure	15%	20%	95%
Wiring-broken wire, shorted wire	5%	0%	70%
Actuators-example starter 70% short circuit/O.L.	30%	0%	60%-90%
20% coil			
10% other			
Sensors-limit switch: 90% mechanical	45%	0%	10%
10% broken wire			
	100%	8%	45%-54%

Source: General Electric Co.

Figure 5.26   Photo of diagnostic I/O. (Courtesy of General Electric.)

hardware faults detected and reported to the CPU, but also diag-
nosed is the health and well-being of the sensors and actuators
connected to the I/O system. Table 5.1 illustrates statistics from
a manufacturer's study of control system diagnostics. This system
check is accomplished by delivering a very short duration voltage
pulse to every output in the system, and then measuring the resulting current in the load. For most loads, the current must be
above a certain minimum low level and below the predetermined
high level. Otherwise, the actuator connected to the programmable controller output is deemed to be either an open circuit or
in a shorted or overloaded condition, and the appropriate status is
reported to the CPU for immediate notification of factor maintenance. This is accomplished through a fully implemented Computer Integrated Manufacturing (CIM) system, or can be used on
occasion on a smaller scale. In either case, the decoding and rapid
delivery of the diagnostic information is critical to the plant's

uptime. The duration of the I/O diagnostic voltage pulse is short enough to prevent most actuators from physically executing. Inputs can be tested in a similar manner but require special treatment externally to ensure minimum current in the normal state. In addition, this technology provides detection and reporting for point level I/O hardware failures. Figure 5.26 shows an example of this new I/O technology.

6
Programming Devices and Alternatives

"A picture tells me at a glance what it takes dozens of pages of a book to expound."

Turgenev

This chapter will examine the wide variety of devices and methods available today to perform basic program creation and editing functions. Both dedicated and personal computer based programming systems, host-computer-based systems, and program documentation will be covered. Programming devices have seen significant changes in the past five years. From an LED-based device, programming devices have evolved to use CRTs as common items, and are now fairly exploding with capabilities to accommodate new and expanded functions. These same programming devices can now become general tools for use in many facets of "factory life." In Chapter 9 we will look at specific techniques for program creation.

6.1 DEDICATED PROGRAMMING DEVICES

Chapter 1 explored the relationship of the programming device to the other parts of the programmable controller system. The pro-

gramming device is used to create the program that the programmable controller CPU will execute, and is occasionally used to aid
in troubleshooting a system that is experiencing a problem. This
has traditionally been the task and scope of capabilities of a dedicated programming device. By its very design, the programmable
controller requires initial programming and configuration to operate. In the early days of programmable controller systems, the
programming device may have consisted of an LED-based device,
with pushbuttons and indicators allowing program creation and
monitoring. Later designs provided for a CRT-based approach, an
important improvement over the LED-based systems. Figure 6.1
shows dedicated programming devices. By this I mean that the
programmer is dedicated to operate with only one brand of programmable controller and is generally limited to the range of
functions it can control. This contrasts to the personal-computer-based devices examined later in the chapter. The dedicated devices have the advantage of being optimized for a particular task,
and therefore, are sometimes more efficient in the execution of
that task than other methods. They automatically have the disadvantage then that the devices cannot be used for anything else
but the narrow scope of functions they were designed for. Most
manufacturers provide a CRT-based device today for the programming of the CPU memory, and some supplement this offering with a hand-held programming device. These devices allow
some of the capabilities provided by their bigger brothers, but
are generally used for small program changes required once the
machine or process has been installed on the factory floor. Figure
6.2 illustrates a hand-held device. The primary method still used
today for programming the systems is ladder logic, as we examined briefly in Chapter 2. This well tested and accepted language
allows a powerful sequence of instructions to be created in a manner that requires little expertise on the part of the user to understand "computers." This is indeed the critical test of any successful programming method used today. The popular software programs designed for use on personal computers today allow an un-
sophisticated user to fully utilize the power of the software without being a master of the "computer." This is indeed the capability of today's ladder logic programming method. From its contact and coil heritage, it always allowed the easy and logical creation of programs designed to control sequential machines and

processes. It has since been embellished with powerful functions allowing data manipulation and number crunching as an integral part of the ladder logic. We will examine this powerful tool in detail in Chapter 9. Figure 6.3 shows an illustration of a typical CRT screen of ladderlogic from a dedicated device.

There is a recent trend that seems to be moving away from dedicated devices to the more flexible approach offered by personal-computer-based systems. The next section details this popular alternative.

6.2 PERSONAL-COMPUTER-BASED PROGRAM DEVELOPMENT SYSTEMS

As noted in the previous section, a strong trend in programming device use positions the personal-computer-based approach in a favorable light. Over the past two years, many manufacturers have announced products that use an industry standard personal computer as the base of their product offering. Some offer the personal computer as part of the product, while many offer just the software to be used with the user's choice of personal computer hardware. Those that offer personal computer hardware as part of their product generally provide a version that has been modified to operate in the relatively harsh environment of the factory floor. This will normally include a computer keyboard that has been sealed to prevent spills and other dirt from entering the area and causing a failure of the sensitive electronics. Also common to this modified design are filter systems to provide positive pressure out of the system protecting the disk drives, and a general unit design allowing operation in extended temperature and humidity conditions similar to that which the programmable controller sees in its normal operation. Disk drives are usually of the encapsulated type, allowing an additional measure of protection on the factory floor. Figure 6.4 illustrates such a personal computer based programming device.

The chief advantage of such an approach is the flexibility achieved from building software tools for a common hardware base. This allows the creation of many varieties of factory floor software that complement each other in different ways. It also allows the efficiency of not duplicating hardware for each solu-

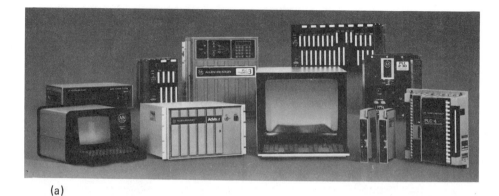

(a)

(b)

Figure 6.1 Photos of dedicated CRT programming devices. (a) Courtesy of Allen Bradley; (b) courtesy of Texas Instruments; (c) courtesy of Square D; (d) courtesy of Cincinnati Milacron.

(c)

(d)
Figure 6.1 (continued)

(a)

Figure 6.2 Photos of handheld programming devices. (a) Courtesy of Eaton Corp.—Cutler Hammer Products; (b) Courtesy of Gould, Inc.

(b)

Figure 6.2 (continued)

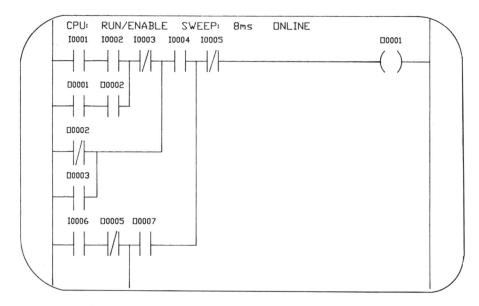

Figure 6.3 Diagram of ladder logic on a CRT screen.

tion is not required. The chief criticism of the dedicated device was that it spent the majority of its life sitting idle since it could only be used for a limited number of functions. In contrast, the personal computer approach can be used in a larger role. We will examine some of the roles the personal computer can take on in its multifaceted design through the use of special purpose software. Included will be the traditional ladder logic programming of course, but also examined will be software approaches for program documentation, data collection and analysis from the programmable controller CPU(s), real-time color graphics operator interface to the CPU(s), simulation of factory floor machines and cells of machines, and network software allowing use as a network administrator or general node workstation.

Programming Software. This is the basic software functionality provided for by most manufacturers as a fundamental requirement. It allows the creation of contacts, both normally open and normally closed, and coils associated with these contacts. In their undocumented form, the contacts and coils will have only the references peculiar to a certain manufacturer's standards, and will have no similarity to anything the user would recognize as applicable to his application. In addition to contacts and coils, this software would allow programming of any special mnemonic functions allowed by the manufacturer in his instruction set.

Documentation Software. This is software used with the program creation software to provide a properly designated road map associated with the application. It allows each contact and coil in the created program to be annotated with an English language name, indicating its unique use in the application program. In addition, a full function documentation program will provide for extensive comments to be added in the middle of the ladder logic, explaining what a particular piece of logic does. More will be said about documentation in the next section.

Data Collection and Analysis. This software functionality is becoming common on personal computers for industrial control. Essentially it consists of the ability to collect data from single or multiple programmable controller CPUs, format it in a predetermined way in a spreadsheet arrangement, and analyze it by manipulating the data in various ways, including bar and pie charts. An example of this is shown in Figure 6.5 It is in this way that production data can be analyzed in a near real-time manner, and "what if" scenarios can be created, playing off various alternative production combinations.

Real-Time Operator Interface Software. This relatively new wrinkle in personal computer software for industrial control allows supervisory control to be established using images on a CRT screen to advise an operator of current process conditions and any alarms that may occur. The operator in turn can input information into the operator interface keyboard (sometimes the personal computer keyboard), to control the process in a supervisory manner. Figure 6.6 shows a typical system using this software. We will see more from this remarkable programmable controller option in Chapter 8 on peripheral equipment.

Simulation Software. This software allows a factory floor personal computer to be used to simulate the flows of a busy factory process or subsystem. Prior to the development of this soft-

(a)

Figure 6.4 Photo of personal computer based programming device. (a) Courtesy of General Electric; (b) courtesy of Gould, Inc.

ware, functionality of this type was available only on very large computer systems. It allows an existing system to be effectively measured and analyzed, and allows process designers to simulate a number of iterations of process designs prior to actually building the process or cell. In this way, throughputs can be accurately estimated and bottlenecks identified early. Figure 6.7 shows a good example of this functionality on a personal computer.

Other Software. As of this writing, the industrial software business is fairly exploding with new applications and options. Included are dedicated functions for a certain speciality such as motion control program development and network management software allowing the personal computer to act as a local area network manager. I am certain the industry will see many more innovative solutions in the form of software for personal com-

(b)

Figure 6.4 (continued)

Figure 6.5 Photo of data collection and analysis software system. (Courtesy of General Electric.)

puters. Even today, personal computers are being used as program development tools with computer numerical controls, robots, vision systems, and a variety of instrument functions for measurement and analysis.

One of the primary benefits of using a personal-computer-based programming and documentation system is the additional ways it can be used, as we have seen above. In addition, since it uses the standard operating system of the personal computer, it can serve as word processor, spreadsheet analyzer, and graphics generation tool with the appropriate software.

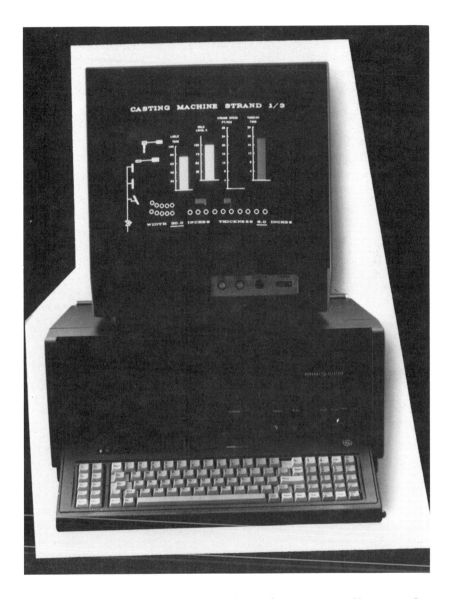

Figure 6.6 Photo of real time colorgraphics software system. (Courtesy of General Electric.)

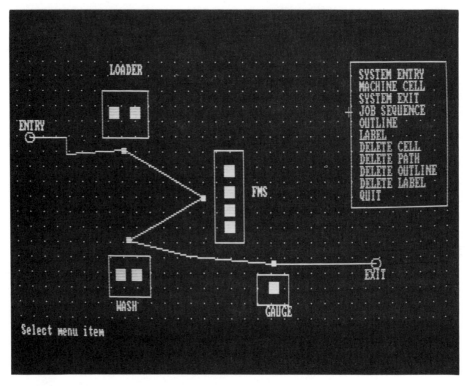

Figure 6.7 Photo of simulation software system. (Courtesy of General Electric.)

6.3 PROGRAM DOCUMENTATION

As noted above, programmable controller program documentation
is an extremely important part of the overall system, and is often
overlooked as such. A properly designed and documented program makes living with the installed system a much more pleasant
prospect than you can expect with no documentation. In spite of
this, many programmable controllers are installed with poor or
nonexistent documentation. As described above, documentation
is a road map allowing the system user to understand just how a
piece of ladder logic software was designed, and what diagnostic
routines to execute if a problem occurs.

A well designed and properly constructed piece of program documentation has several segments, each with a specific purpose, but each complementing the other in an overall plan. First we will consider the ladder logic itself. When the program idea is first conceived, the programmer may begin with a sketch of some of the logic required, or he may just plunge into the task on the program development terminal. With today's programming tools, this is a perfectly acceptable alternative because the documentation software is usually combined with the program creation software in a personal computer. Even where this is not the case, there are many stand-alone systems designed to perform documentation tasks after the undocumented ladder logic program is complete. In any event the programmer begins his task in a more or less formal fashion. Good documentation practice calls for each contact and coil in the ladder network to be labeled or "annotated" with an English language notation, indicating just what function that contact or coil has. This is generally done with a shorthand nickname, and is later expanded into a more explicit reference after the documentation process is further along. The nickname may consist of a single word of say, seven characters, while the expanded reference can be up to three lines of single seven-character words in some systems. Figure 6.8 illustrates an example of this contact and coil annotation. An important part of the documented ladder logic is the text or "comments" inserted strategically among the rungs of ladder logic. These comments give additional meaning to the annotated contacts and coils by describing exactly how a particular section of logic is designed to operate. It may even contain suggestions of what to do in the event a problem occurs, such as what coils to force, or what external sensor or actuator to inspect. This can save a tremendous amount of time and aggravation, especially when a critical production system is shutdown, and every minute of downtime is costing hundreds of thousands of dollars of lost or ruined production. Beyond the contact and coil annotation, and the comments inserted in the logic, a third feature is critical to a well-documented system, and is available in many commercially available systems today. That is the ability to generate what is called "cross references." Cross reference features in a ladder logic diagram are automatically generated in those systems so equipped, and provide an easy method to determine the sometimes complex interrelationship between a

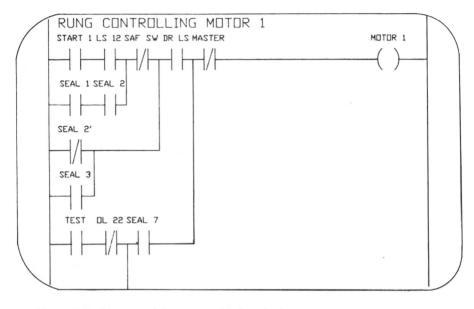

Figure 6.8 Diagram of documented ladder logic.

given coil address and the contacts that share the same address and reference. Since a programmable controller allows flexible coil and contact assignments, a typical coil can have anywhere from one up to fifteen or more associated contacts scattered throughout the logic. This can be extremely annoying, especially when other parts of the documentation are poorly done. The cross referencing feature of documentation software automatically scans and prints alongside the coil a series of rung numbers (or other designators), indicating where the associated contacts can be found. Another feature generally found in good documentation packages is the printing of register and I/O status at a given point in time, along with a list of I/O and register references used in a program. This allows any subsequent work done on the software to avoid conflicting with existing references.

The systems we have been looking at can come in two basic forms: (1) a dedicated system (Figure 6.9), designed solely for the purpose of program documentation, and (2) a personal-computer-based system, which combines the program creation and

Figure 6.9 Photo of dedicated documentation system. (Courtesy of Process and Instrumentation Design Inc.)

documentation into one tool. When the latter system is used, connected on-line to the programmable controller CPU, it allows real-time access to the process using a highly intelligent "documented" window. The final documented logic program can be viewed on a CRT, or can be printed onto a printer. This automates a process that normally required substantial drafting resources and was performed manually. Once final documentation is created, checked, and approved, multiple copies should be made for use by maintenance personnel on all shifts. Any subsequent changes should be regenerated, replacing all of the existing copies.

6.4 HOST-COMPUTER-BASED SYSTEMS

The term host computer has come to mean a great many things in the last few years. For our purposes, we consider a host computer to be a multi-user system, normally centrally located in a facility, that is used for a variety of tasks. These tasks can range from predominantly manufacturing in nature, such as shop floor scheduling, maintenance and tool management, and MRP, to broader tasks that include accounting and other financial functions.

Now that we have defined a host computer for our purposes, let's examine how it is used as a programming and documentation tool in a modern factory system. In a factory that has accumulated a large (100 and up) number of installed programmable controllers, it is quite often cumbersome and impractical to support all of these with portable programming devices. The variety of installed CPU's along with the expense of purchasing and supporting a number of programming devices cause many facilities to desire a method by which the program development and maintenance for these programmable controllers can be done on a centralized and common basis. To address this need, a number of firms have developed software and communications products that allow program development and documentation from a central location, using a terminal connected to the plant's computer. Sometimes, depending on the nature of the host system, the terminal may be simply a "dumb" terminal, or can be an intelligent terminal as part of an engineering design workstation. Either way the process is basically the same.

The software that is resident on the host computer is designed to allow ladder logic program development and documentation. The perspective that a programmer has is essentially the same as if he were using a portable personal computer connected directly to the programmable controller CPU. The programmer reacts to the terminal screen with the keystrokes required to create, edit, or document a piece of ladder logic software. When complete, the final version of the software is transferred to computer memory, and can then be directed to any CPU the system desires over the plant communications network. A program may be extracted from a programmable controller on the factory floor for examination or editing, and then can be transferred back to the appropriate CPU. Depending on the plant size and age, this network may be a simple baseband system, or may be a multichannel broadband system. The latter style is that advocated by the IEEE 802.4 Token Bus Standard, commonly known as the Manufacturing Automation Protocol (MAP) Standard. This standard was originally nurtured by General Motors, but has since been endorsed and adopted by a large number of discrete parts and process industry participants. We will take another look at the MAP system in Chapter 12. Figure 6.10 shows the diagram of

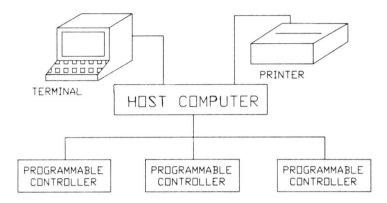

Figure 6.10 Diagram of host computer-based programming system.

a system of host computer utilized for programmable controller programming documentation.

While the number of existing systems today that do programmable controller program documentation and edit on a centralized basis is small, it is a growing trend. As the cost of communication hardware and software decreases, and standards on networking become solidified, we will see more and more of these systems. This is especially true in plants that have a high installed base of existing or planned programmable controllers. One of the more interesting trends is that of using the programming device as a node on the network for both program access, and process monitoring and control.

7
System Configuration

The bits and pieces of programmable controller equipment, while interesting, only solve parts of the problem. System configuration—the assembling of the bits and pieces into a workable form— begins the process of turning the "apparatus" into a "solution."

This chapter will look at the task of bringing together the pieces of the programmable controller system into a unit designed to solve a specific problem. Included will be the various CPU and I/O racks, along with their associated power supplies and cables. Basic guidelines will be examined regarding initial system checkout, once assembly has been accomplished. The chapter is not intended to replace the installation and service manuals of the programmable controller manufacturer, but is designed to give an overview of the system configuration process in general. This process is then followed by the very important step of developing and integrating the application program software. This is covered in Chapter 9.

7.1 CPU TO I/O RACK CONFIGURATION

Depending on the particular system provided by a manufacturer, there may or may not be a separate CPU from the I/O system. This is generally a function of the system size, with smaller systems not requiring separate I/O because it is designed as an integral part of the system along with CPU and power supply. Figure 7.1 illustrates several examples of each design. Some of the programmable controller systems available today incorporate some I/O in the CPU chassis, and allow expansion to other chassis that serve as I/O only chassis.

Cables are used to connect the CPU chassis to the I/O chassis. These can be either parallel communication cables, employing sixteen twisted pair of conductors, for example; or can be serial in design, using perhaps two twisted pair of conductors. The choice is usually made between a trade-off of distance versus speed, with the parallel design giving better communication speed performance, while the serial option delivers satisfactory speed at a manageable cost for cable and installation. Both options are not available from all manufacturers, with some systems only offered with the serial option. Connectors are provided on many cables sold by the programmable controller manufacturers, especially the parallel communication cables. These connectors are normally designed to a standard, such as the 25- or 39-pin D type connectors. Cables with connectors assembled and tested are normally available in increments of length, from 2 to 500 feet. In the case of serial cables, the manufacturer's documentation usually publishes the cable conductor specification, and refers the purchaser to a cable vendor. The use of these cables sometimes presents special challenges. For example, the pulling of a cable through the conduit it is to reside in requires some planning. The size of the conduit, along with the numbers and types of cables to share the conduit, are important considerations. Also important to consider is whether any connectors already installed on the cable will be able to make the installation trip through the conduit.

Whether parallel or serial, the CPU must provide the means for the I/O to be communicated with. In the case of the parallel option, this normally involves an I/O control module that resides in the CPU or first chassis that drives a single or multiple chassis. The serial option also requires a driver module, but can normally

be arranged a bit more flexibly than the parallel choice. Depending on the manufacturers design, the system can be configured
with I/O in a daisy chain fashion, or can be arranged in a star connection. Figure 7.2 shows the various I/O configurations as they
relate to the CPU and I/O driver modules.

As noted in an earlier chapter, a power supply is required to
provide adequate current for the electronics on the resident circuit boards. In addition, some designs call for the system power
supply to have additional capacity to handle external loads as
well. The power supply can be external or internal to the CPU
or I/O chassis, although the trend today is to make the power
supply an internal, modular design. This allows servicing of the
supply in the same manner that an individual circuit board would
be. The choice of supply capacity, if there is an option, is made
on the number of modules to be installed in a chassis, and the
current requirements of each. Most programmable controller
manufacturers today provide a standard capacity power supply
and a high capacity model. The latter is generally required where
a large number of high-current-using modules are to be installed
in a single chassis. Examples of high-current modules are analog
modules, and most intelligent modules. If a system configuration appears to be challenging the power supply capacity, an alternative is to redistribute the modules among several chassis so
that a balance is achieved between high- and low-current modules.
Figure 7.3 shows examples of programmable controller systems
and subsystems using both internal and external power supplies.

7.2 I/O RACK CONFIGURATION AND WIRING

Depending on the design of programmable controller system that a
manufacturer has to offer, the configuration of the I/O rack or
chassis can be similar to that of the CPU, or it can be quite different. For example, some CPUs are just that, a CPU *only*, with
I/O totally contained in other external chassis. Others have a
minimum of I/O installed in the CPU chassis, and then expand to
other I/O only chassis. For purposes of this section, we will consider the I/O-only condition.

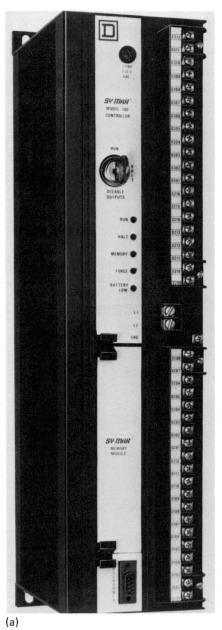

(a)

Figure 7.1 Systems with and without integral I/O. (a) Courtesy of Square D; (b) courtesy of General Electric.

(b)

Figure 7.1 (continued)

In an earlier section, we saw where a CPU sends out I/O communication to a single or multiple I/O chassis in a parallel or serial
fashion. Each I/O chassis that is to receive that communication
must be properly outfitted with the hardware to accomplish the
task. This requires an I/O receiver module to be installed and
properly configured in the target chassis. Depending on the design
of the programmable controller and choices of I/O flexibility, the
module can receive parallel information from a sixteen-pair cable,
or serial information from a two-pair cable. Some of the newer
technologies in I/O build in the I/O receiving functions as part of
the I/O block, eliminating the traditional receivers, chassis, and
power supplies. Figure 7.4 shows an example. In any event, the
function of the receiver is to receive information in a rapid manner, decode it with high integrity, and deliver it to the appropriate
I/O module resident in the chassis. It also acts in the reverse, of
course, collecting input information from the input points in a

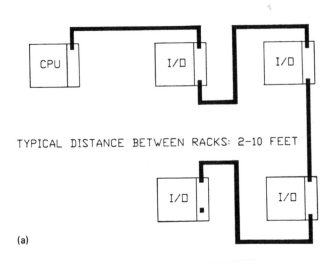

TYPICAL DISTANCE BETWEEN RACKS: 2–10 FEET

(a)

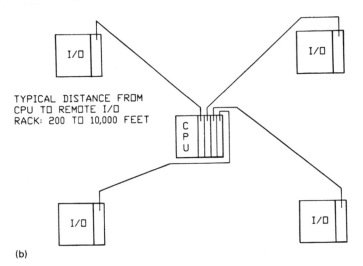

TYPICAL DISTANCE FROM
CPU TO REMOTE I/O
RACK: 200 TO 10,000 FEET

(b)

Figure 7.2 Diagrams of parallel and serial I/O systems. (a) Parallel; (b) serial.

chassis, encoding them as serial or parallel information packets, and making them available for transmission to the CPU. In addition to the I/O control information supplied by the receivers, certain models provide selectable functions for I/O system diagnostics. For instance, a jumper selectable function in some designs provide for predetermined conditions to occur if power fails or I/O communications malfunction. This will normally call for a chassis to maintain last state in all of its outputs in a failure condition, or for the chassis "downstream" from a chassis that malfunctions to continue to operate as normal. In both cases, the failure mode information is communicated back to the CPU, which can take appropriate action as determined in current logic.

Small programmable controller systems present no major challenge in installation, insofar as panel space or wiring is concerned. Medium and large systems, however, pose some special challenges for the system installer and integrator. A large system can have 30 to 50 chassis connected together, each with 10 to 16 I/O modules inserted in it. The first choice after the initial layout of I/O is made is to decide whether the chassis will be rack mounted or panel mounted. Rack mounting usually is done in a control room situation, where there are 19-inch racks available, and other conditions favor rack mounting. Panel mounting, on the other hand is generally done so that programmable controller CPU and I/O racks can be installed in a closed door cabinet on the factory floor. In both cases, the power and I/O wiring is accomplished by running individual conductors to the appropriate connectors.

Cable trays mounted to the front of the I/O racks provide a method to direct and control the large number of conductors in a typical installation. I/O communications is accomplished by pulling and terminating the communication cables referred to earlier. I/O module placement in the rack is determined by a number of factors. First, as noted earlier, the rack power supply may support only so many of a certain type of module, and reassignment of module placements may be necessary to redistribute the current loading. Secondly, the addressing of each slot in the chassis may be flexible or rigid, depending on the manufacturer's design. This fact alone may cause considerable difficulty in system configuration, since the unexpected redistribution of modules in rigid system requires that extensive changes be made in the lad-

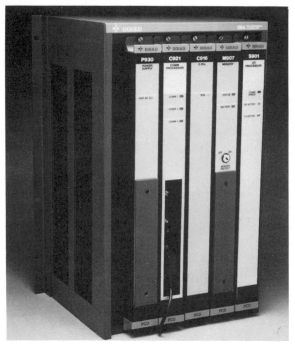

(a)

(b)

Figure 7.3 Photos with internal and external power supplies. (a) Courtesy of Gould Inc.; (b) courtesy of General Electric; (c) courtesy of Gould; (d) courtesy of Allen Bradley.

(c)

(d)

Figure 7.3 (continued)

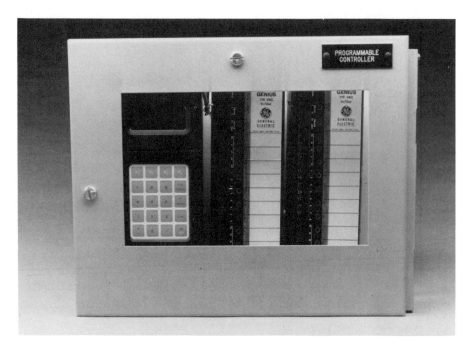

Figure 7.4 Photo of I/O system with integral I/O receiver. (Courtesy of General Electric.)

der logic software to accommodate the change. Figure 7.5 illustrates I/O chassis with a variety of modules installed. As an interesting observation, the current trend in today's system is to distribute the I/O to greater lengths and depths than ever before. This will present special system configuration challenges in that this distributed I/O is likely to be mounted in unusual places, at least by today's standards. Documentation and system diagnostics will play a key role in the system of tomorrow.

7.3　INITIAL SYSTEM CHECKOUT

It is assumed that the individual pieces of the programmable controller system will have been examined closely by this time, and

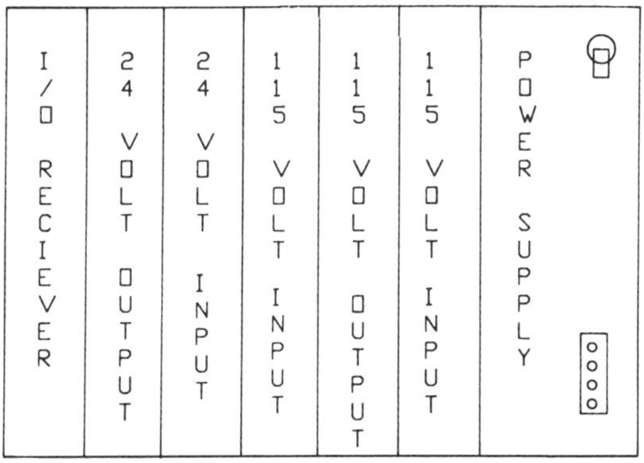

I/O RECEIVER	24 VOLT OUTPUT	24 VOLT INPUT	115 VOLT INPUT	115 VOLT OUTPUT	115 VOLT INPUT	POWER SUPPLY

Figure 7.5 Diagram of chassis with I/O modules installed.

the list of material compared to that of the purchase order. As a practical matter, the appropriate serial numbers of the purchased equipment should be recorded and placed in a safe place along with the system documentation and manuals for both installation, programming,and maintenance. Generally, the CPU will have a unique serial number, and occasionally, other components will also have unique numbers that should be recorded. These numbers will be important should any service work be required, whether in or out of warranty. Hardware and software upgrades are often performed on the basis of serial number verification. It is therefore critical that the manufacturer has the system recorded in warranty records to insure that notification occurs.

At this time, the power connected to the system should be examined and verified for correct wiring per the manufacturer's recommendations. This will generally include connections of AC voltage, L1 and L2 to the appropriate terminals. Care should be taken prior to connection that the correct voltage connections are made; that is, many systems are selectable for 115 volts or 230 volts at the power supply. In addition, the power supply may have terminal connections for an external battery to back up the on-board memory battery system, and dry contact alarm connections which are used to connect external alarm mechanisms to signal if a CPU has stopped operating unexpectedly. All of these connections should be checked visually, and for continuity with a suitable instrument.

I/O wiring should be checked for continuity at this time in a manner similar to that used above. Before power is applied to the loads connected to the outputs of the programmable controller, it is possible and sometimes recommended that an I/O wiring checkout be performed. This can be done by connecting the programming device and performing some basic troubleshooting exercises on the system. By utilizing the "force and toggle" commands from the programming device keyboard, the user or system integrator can individually exercise each input and output in a controlled manner, observing the resulting action on the output circuits. It is especially convenient if the programming device can be connected at the remote I/O point, although some systems require that the programming device be connected at the CPU only.

If all is well on the I/O wiring, power wiring can now be connected to the output circuits. The CPU and I/O racks can then be powered up and observed for proper operation. An option at this point is to run the system in a disabled mode. That is, the CPU takes input information, solves logic, and updates the output status table, but all outputs are disabled. This is sometimes used as a final check before fully operating the system. With all preliminary checks accomplished, the system can now be declared operational, and any loading or modifying of ladder logic software can begin.

8
Peripheral Equipment

A programmable controller system without the benefit of peripheral equipment is like a very smart man in a dark, soundproof room.

Chapter 8 is entitled peripheral equipment, but might be more descriptively called the "windows to the world and other intelligent devices." In this chapter we will examine the next layer of the entire control system beyond the CPU, I/O, real-world devices, and programming devices. Questions of interest include: How the programmable controller articulates to human operators in both temporary and permanent forms; how it provides records of its program storage, how the primary CPU can be backed up in a critical application, and how the system can be integrated with special purpose control equipment.

8.1 MAN/MACHINE INTERFACE TECHNIQUES

The term man/machine interface is used here in its broadest form. It could also be referred to as operator interface, or system control station, or any of a number of other terms. The basic notion is that, in order for a human to interact with the programmable control system during operation, some effort is required on the part of additional hardware (and sometimes software) to accomplish this task. Depending on the application requirement and industry practices, the man/machine interface can consist of a few pushbuttons and indicator lights, all the way up to an intelligent color graphics terminal residing as a node on a local area network. We will define three levels of need in the man/machine area.

Level 1. This is the most basic and common type of man/machine interface, and involves the selection and wiring of pushbuttons, indicator lights, and other discrete devices that are mounted on a panel to designated inputs and outputs of the programmable controller system. With appropriate CPU ladder logic programming, these devices can take commands from, and report status to the human operator. Figure 8.1 shows a machine with this type of pushbutton panel interface.

Level 2. A factory hardened CRT-type of interface comes into play here, and while it can be monochrome or color, it is normally configures as the "dumb" or "semi-intelligent" type. That is, all, or a large part, of the information displayed on the terminal comes from outside the terminal through the communications interface. In some cases "static" graphics screens are stored in the "semi-intelligent" version of the terminal, and the only data coming from the outside is variable data to be displayed within the graphics. Factory hardened in this case means that the terminal is usually designed to be gasket mounted in a panel or cabinet door right on the factory floor. With this installation, the device can then be considered suitable for hose down or other NEMA 4 conditions. Figure 8.2 shows an example of this type of CRT interface.

Level 3. This is the highest and most complex form of man/machine interface used with programmable controllers today. It involves the use of an "intelligent" terminal, many times with color attributes to more accurately and quickly communicate more information to the human operating the equipment. This interface will use its own microprocessor and memory separate from the programmable controller. It can be connected directly to the programmable controller through a communications interface, or can reside as a node on a local area network. Figure 8.3

Figure 8.1 Photo of pushbutton and indicator panel. (Courtesy of General Electric.)

shows a color graphics operator interface system of the level 3 type.

All applications requiring man/machine interface are not clearly definable as level 1, 2, or 3. Indeed, there are times when a combination of two or more levels are appropriate. This can occur because of some safety related design requirement, or because a CRT cannot convey enough information fast enough. Figure 8.4 shows such an example of combined attributes.

As one might imagine, the development and use of these levels of operator interface devices was an evolution over a period of time. For most of the 1970s and even into the early 1980s, the primary method of man/machine interface was that described in level 1. A CRT interface was considered an impractical luxury except on the most complex and extravagant systems. The CRT

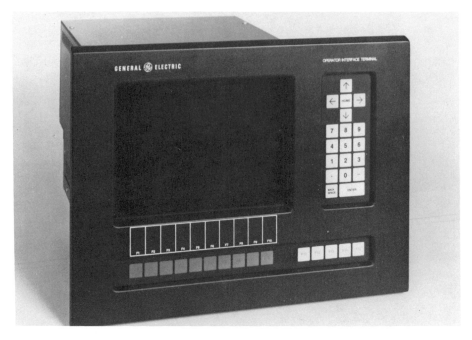

Figure 8.2 Photo of factor hardened CRT. (Courtesy of General Electric.)

programming device was the first of its type for most users, and was normally dedicated to programming only, not machine interface. The process industries were the drivers for the next major step in man/machine interface development. Both the continuous and batch industries were using color CRT-based systems by the early 1980s for rapid communication to their operators of process malfunctions or alarms. These early level 3 type systems gave way to the stand-alone intelligent terminals available today for use with most programmable controllers. Discrete part manufacturing industries soon found many applications for intelligent systems as well. Oddly enough, the level 2 type solution was the last to enjoy broad development in the industry. Perhaps this was because the factory hardened CRT was difficult to produce economically in a design suitable for direct machine mounting on the factory floor. This was the solution required by many ma-

Figure 8.3 Photo of intelligent color terminal. (Courtesy of Metra Inc.)

chine builders and system integrators. Figure 8.5 shows a diagram of how the different levels of man/machine interface might be used in a system example.

Some examples of recent applications of level 3 operator interface cover both process and discrete part manufacturing. Process control is more traditionally expected in that industry, as their need for timely and broad distributed control was well known and understood for some time. It involves the supervisory control of many process loops (defined in the next section), and required a friendly way for operators of varied skills to control the process. The discrete parts industry application evolution was not so clear, however. As pioneers tried different solutions, it

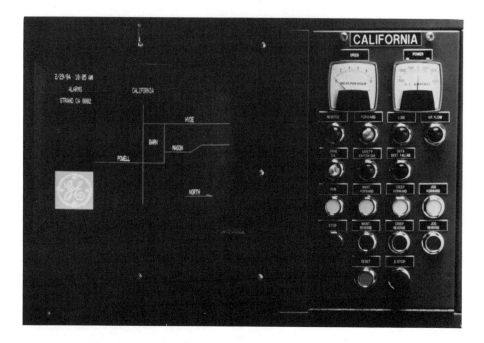

Figure 8.4 Photo of combination pushbutton and CRT panel. (Courtesy of General Electric.)

quickly became obvious that this industry segment had many opportunities to use the level 3 solutions; including the need to monitor and control large material handling oriented applications. Figures 8.6 and 8.7 illustrate process and discrete part application examples.

8.2 PID LOOP CONTROLLERS AND THEIR INTEGRATION WITH PROGRAMMABLE CONTROLLERS

When we first examined the needs and applications of PID (Proportional-Integral-Derivative) control in Chapter 5, we were generally referring to a special purpose I/O module residing in the I/O chassis, or logic performed in the CPU and executed through

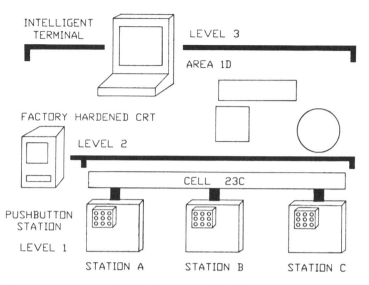

Figure 8.5 Diagram of 3 level operator interface system.

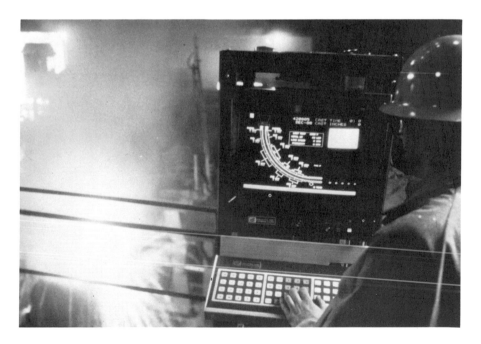

Figure 8.6 Photo of process example operator interface. (Courtesy of Industrial Data Terminals Corp.)

Figure 8.7 Photo of discrete parts example operator interface. (Courtesy of Texas Instruments Inc.)

analog I/O modules. In this section we will look in detail at the powerful cousin of this approach, the stand-alone loop controller. These electronic devices are not new, but have been used for many years in the continuous- and batch-process control industries. In early times they were of an analog design, which coincidentally fit the physical parameters they controlled such as temperature and pressure. In more recent designs, they, like many electronic systems, have taken advantage of the power of the microprocessor to accomplish accurate closed-loop control of these physical parameters. Figure 8.8 shows a typical stand-alone digital PID loop controller with hand-held programming terminal. These loop controllers are generally available in either single-loop or multiple-loop configurations. A multiple-loop rack-mounted configuration is shown in Figure 8.9. They allow an operator to establish the process setpoint from the front panel of the loop controller operator interface. They also allow monitoring of the process variable, alarams, and output, and permit the process to be run "manually" by the operator if necessary. Figure 8.10 illustrates a close-up photo of a typical PID loop controller operator panel.

It is the digital nature of the newer loop controllers that has allowed the next step of control system integration to occur. Mi-

Figure 8.8 Photo of PID loop controller. (Courtesy of General Electric.)

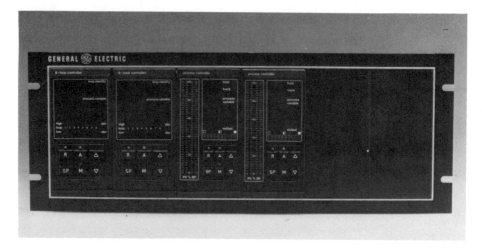

Figure 8.9 Photo of multiple PID loop system. (Courtesy of General Electric.)

croprocessor designs have allowed the loop controller to communicate efficiently with external devices intelligently. In our case, the intelligent device is the programmable controller. The results of this integration provide process control engineers with (1) the ability to have local independent control at the process (loop controller) level, (2) the functionality of communication of setpoint and process status data to the programmable controller, and (3) efficient integration of sequentially oriented discrete control of pump motors and the like through normal programmable controller I/O. The communication to the loop controllers is generally handled in a serial fashion from a special purpose ASCII I/O module in the programmable controller system chassis. A system configured in this fashion can normally communicate to multiple loop controllers. Figure 8.11 shows a diagram of such a system, and Figure 8.12 shows loop controllers with a suitable programmable controller for configuring such a system. It is interesting to note that such a system, especially if it is a large one, may use an intelligent color operator console of the level 3 type in a control room installation. This would allow an operator or supervisor to monitor and execute setpoint control on the loops integrated with the programmable controller. Such a console is shown in Figure 8.13.

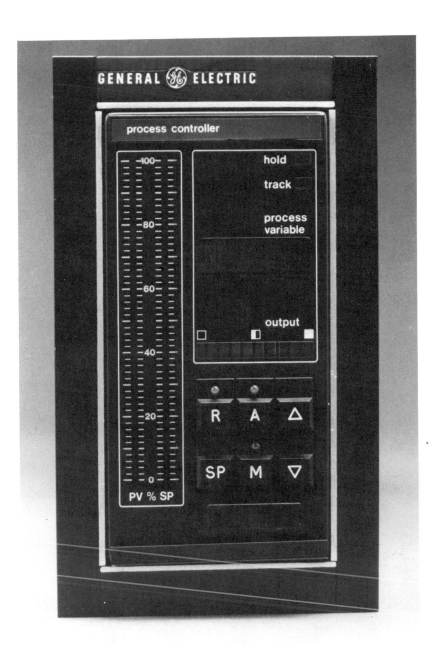

Figure 8.10 Photo of PID panel. (Courtesy of General Electric.)

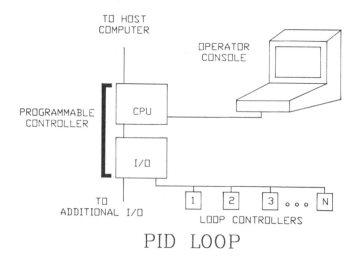

PID LOOP

Figure 8.11 Diagram of programmable controller, loop control, and operator interface system.

Figure 8.12 Photo of programmable controller with loop controllers. (Courtesy of General Electric.)

142

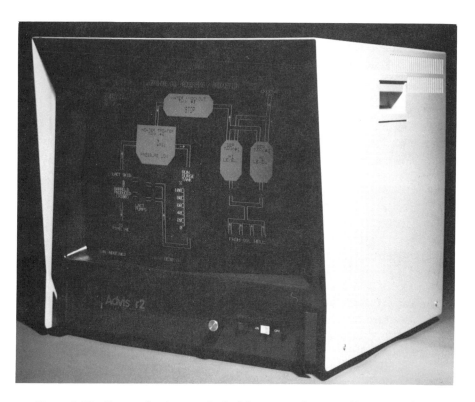

Figure 8.13 Photo of color terminal with process diagram. (Courtesy of Allen Bradley.)

Applications for loop controllers are broad and varied, and are sure to broaden even further considering the new flexibility their integration with programmable controllers brings. Some examples of batch process control suitable for use with loop controllers are heat treat furnace control, plastic injection molding control, and batch weighing system control. Continuous process control applications are even more broad and include many which are fluid related, such as chemical and petroleum, and web related such as paper, rubber, and steel manufacturing.

8.3   DATA STORAGE AND RETRIEVAL

In Chapter 6 we saw how the programming device is used to create
and edit a ladder logic program to be used with a programmable
controller. An important element of that and other related applications requires the use of peripheral devices for storing the information for later retrieval and use. This is generally accomplished
by using magnetic storage systems, either as a separate device or
integrated into the programmable controller programming device.

Magnetic storage systems are currently available in two general types: magnetic tape based and magnetic disk based. Both
use a coated media to allow information to be encoded onto the
media in magnetic form, and later detected and decoded into useful information. Tape systems use a media very similar to audio
grade, on two reels arranged in a cartridge configuration. Tape
systems are commercially available as a stand-alone or suitcase-
based unit, or can be integrated by the programming device
manufacturer into the device. This is especially useful and important on those programming devices designed to perform as off-
line program development terminals. Ladder logic programs developed in this type of device can be stored routinely on the magnetic tape system.

Disk based systems, on the other hand, are newer and use a
flat magnetic media, encapsulated in a variety of ways to protect
the fragile media from foreign material. This type of storage
media has become more common in the programmable controller
industry with the use of personal computers as program development devices. Functionally, these systems operate similar to the
tape systems, but are less commonly found as a stand-alone system. They are most often found as part of the programming device.

In addition to finding use as a storage system for programming information, these systems are often used as process information storage devices out on the factory floor. They can be
connected to specific intelligent devices, such as the programmable
controllers running the process, or in some cases, they might reside as a node on a local area network (LAN), acting as a common
repository of information for both storage and retrieval use. This
is especially common on processes that are regulated by agencies
of the Federal Government that require historical process informa-

tion for verification purposes. Information gathered in this way
is usually analyzed in an off-line fashion. Figure 8.14 shows a
commercially available tape unit, while Figure 8.15 shows a family
of programmable controllers along with a programming device
utilizing a disk based storage system integrated in the device.

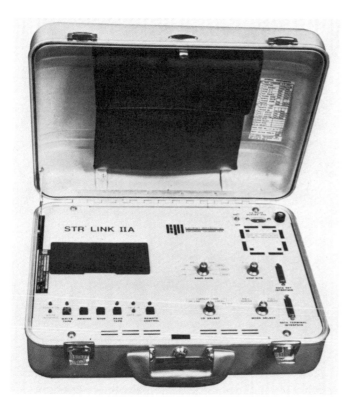

Figure 8.14 Photo of portable tape unit. (Courtesy of Cincinnati Milacron.)

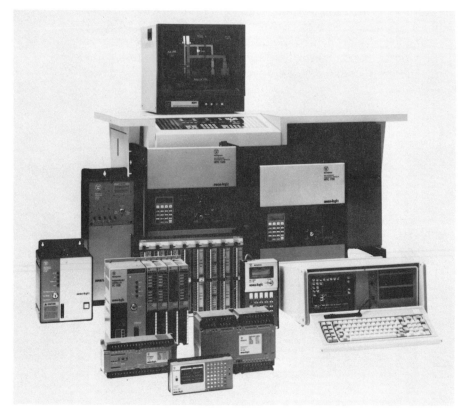

Figure 8.15 Photo of disk based programmer with programmable controller family. (Courtesy of Westinghouse.)

8.4 REDUNDANCY AND HOT BACKUP SYSTEMS

There is a growing category of programmable controller applications today that require, or could benefit from, a system configuration that provides increased system availability through some form of redundancy or hot backup. Redundancy is defined in most dictionaries as "superfluous repetition." While that is accurate in our case as well, it is not necessarily descriptive enough. Redundancy in programmable controller systems provides that all or critical parts of the system have identical "twins" on-line ready to operate at the failure of the other twin. This involves at a minimum the duplication of the CPU function, and can include the I/O in some manufacturers' systems. Depending on the system design and the criticality of the application, the redundancy option can include a separate device (sometimes referred to as a redundant processing unit) that acts as a synchronizer of the two duplicate CPUs, and rapidly switches control from the master to the standby in the event it detects a failure. Less complex systems use two CPUs wired together on a communication link, with the master signalling the standby in case of critical failure. This alternative does not normally provide the rapid rate of transfer available by using a redundant processing unit. This simpler, less rapidly responsive system is sometimes called a hot backup system. Figure 8.16 shows the difference between the two approaches, and illustrates the single and dual I/O chain option as well. As a practical matter, some installations may not require the level of redundancy described here, and as an option, keep a standby CPU configured but not powered, ready to be mounted and cabled in the event of a failure of the installed system.

The concept of redundancy in control is not a new one. Most of the control systems designed for the process industries have for some years offered redundant processing at some level. These critical applications in the chemical, petroleum, rubber, paper, metals, and other industries called for the absolute ability to maximize system availability to its highest possible limit, as a control system failure generally means a high financial penalty from lost production or damaged equipment. So it should come as no surprise that the process industries drove the development of programmble controller systems with redundancy options. And while it is true that most programmable controller applications in process industries are for ancillary functions, the redundancy requirements are just as important as the core process control. Most major manufacturers of programmable controller systems today offer a redundant or hot backup option. Some of the newer entries offer controllers that are designed especially for double or triple redundant applications. An example of such a system is shown in Figure 8.17. While the process industries are still today the primary driver of redundant control applications, the discrete part industries are also finding a larger number of control applications that can benefit from improved control

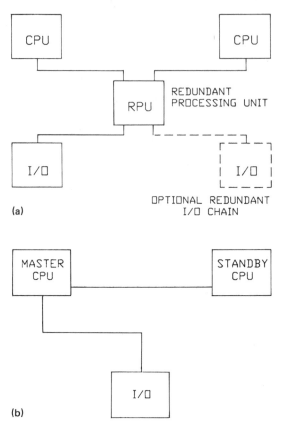

Figure 8.16 Diagrams of (a) redundant and (b) hot backup systems.

system availability. Included in this group are synchronous metal cutting and forming processes, along with high value assembly systems such as an automotive assembly where a single failure can stop a high cost process where the lost opportunity cost can be $5,000 a minute.

A typical sequence of functions that would accompany a failure in a redundant control system would be as follows (refer to Figure 8.16): Prior to the failure, the master CPU is in control, with the standby CPU being synchronized and updated with I/O and register data from the master by the redundant processing

Figure 8.17 Photo of triple redundant system. (Courtesy of Triconex, Inc.)

unit. The redundant processing unit suddenly detects a signal from the master CPU indicating a critical failure of its ability to perform as required. Within one to three scan delays, the redundant processing unit transfers control from the master CPU to the standby CPU, making it the new master, with the new standby being signalled for off-line repair. For critical systems of this type, a measure of time, called mean time to repair (MTTR), becomes critical as it can limit the overall system efficiency by not providing a CPU that is quickly serviceable to return to on-line status as standby CPU.

As a matter of interest, it is important to recall from Chapter 5 that most system failures come not from the programmable con-

Figure 8.18 Photo of printer with process terminal. (Courtesy of Square D.)

troller system, but from the sensors and actuators attached to it.
And while dual I/O chains offer some measure of improvement,
preventive diagnostics integral to the I/O apparatus as described
in Chapter 5, along with redundant CPUs, provide the best available combination of system integrity and availability.

8.5 PRINTERS AND OTHER HARD COPY PERIPHERALS

As part of modern control systems, printers are not a new item.
They have been used traditionally as tools to provide a hard record
of process events that occurred. This is especially important in
processes that are regulated or inspected by government agencies,
such as food and pharmaceutical. Another example are some
exotic metal treatment processes where large amounts of information is collected to verify the integrity of a batch of processed
material. Another application that finds more and more printers

in use is the recording of process alarms and their acknowledgment at the time they occurred. Part of this information is presented on the operator's CRT screen, while a permanent record is created on the printer. The proliferation of low cost printers, along with the availability of microprocessor-based devices to drive information to them, accounts for the higher rate of use in industrial applications. Industrialized versions of these printers are becoming available which provide a higher level of protection from foreign material harming the printer. Figure 8.18 shows a printer as it might be used as a logging device for a process control application.

9
Programming Languages and Techniques

Like the CPU and I/O systems, programming languages and techniques join the ranks of critical, yet distinctive parts of the programmable controller system. Without them, the system would appear quite a mysterious device.

As with any other endeavor, programmable controller language articulation is key to successful execution. Language design defines the ease or difficulty with which the control engineer creates the control program. This chapter examines some of the more common methods used with programmable controllers today, along with some of the more obscure. Included are relay ladder (perhaps the most common), function block, boolean, and a category defined as special application programming. The chapter concludes with an introduction to programming training options generally available today.

Regardless of the language specified or used by the programmable controller manufacturer, the design, of both the language and programming device and the software that controls it, must be carefully thought out and tested to ensure the easy use by a variety of end users and system integrators. Indeed, a larger portion of development programs is dedicated to programming software development each year by the manufacturers. This is especially true with the more prevalent use of personal-computer-based programming devices. A properly designed programming system should allow the user with modest skills and limited computer sophistication to become comfortable with the system in 2 to 3 days use. It should make extensive use of English (or other), language menus, and should provide excellent written documentation supplemented by on-line help text(s).

9.1 RELAY LADDER PROGRAMMING

Relay ladder programming, sometimes called relay ladder logic, is one of the most widely used programming methods today. It has evolved (some would say survived) from the control-relay based systems of previous years. The main reasons that it has transcended its hardwired heritage with only minor modifications are (1) that most factory floor electricians who are responsible for maintaining programmable controller systems are most familiar with relay ladder logic, and (2) that in spite of so-called higher-order language developments, few have been found to so completely cover the needs of near-real time control including the representation of input and output point status. Even with the increased use of math and other data intensive instructions in the ladder structure, the overall acceptance and preference of relay logic continues. This is sure to change, albeit slowly, as more of the general population becomes computer-comfortable and language developments continue.

As we noted in earlier text, relay ladder logic is so named because it uses two symbolic vertical "rails" and forms "rungs" from the contact and coil networks. Practice calls for an assumed voltage of 115 volts across the vertical rails, with circuits completed by specific logical circuit completions from the transitions of contacts. Let's look at a specific example. Figure 9.1 shows a diagram of a typical simple logic network. Imagine that 115 volts are "measurable" across the two uprights. The normally open contact labeled START is the representation of a real-world pushbutton, as is the normally closed contact labeled STOP. The symbol at the right represents an electrical coil, and in this case is labeled MTR, perhaps as a motor starter. A normally open contact, labeled MTR, is shown "wired" back to the front of the

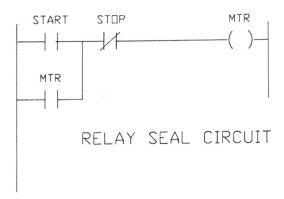

Figure 9.1 Diagram of relay seal logic example.

network, in parallel with the start contact. This can be envisioned to represent a contact from the motor starter MTR and we will see how it is used in a moment. Remembering that current flows through a normally closed contact in its normal state and a normally open contact in its "abnormal" or transitional state lets see how this circuit operates. Assuming the MTR coil is off when we begin, we first push the start pushbutton, which transitions that contact in the programmable controller ladder logic, and allows "current" (or power) to flow along the rung. Power continues through the stop contact, since it's in its normal condition, and arrives at the "MTR" coil, energizing it. This is no different than turning the switch on a table lamp. Now let's see what happens to the contact labeled MTR at this time. With the MTR coil energized, power flows through both the start pushbutton and the MTR contact. It should follow, therefore, that the start pushbutton can now transition to "open" again and power will continue to flow from the MTR contact, through the stop pushbutton contact to the MTR coil, keeping it energized, without pushing the start pushbutton continuously. This is commonly called a "seal" circuit, since the MTR coil is "sealed" in by the MTR contact. In fact, the coil will continue to be energized until the stop pushbutton is pressed (or system power is removed), momentarily opening the circuit, and breaking the "seal." It is sometimes called a "latch" circuit, although this is an error since a latch would maintain its status through a power outage while a seal will not.

The preceding example illustrates basic relay logic, in ladder form—hence the term "ladder logic." Most programmable controller instruction sets can be broadly separated into two segments. These are referred to in a variety of ways, but can generally be called the basic and enhanced instruction set segments. Table 9.1 shows a list of what might be included in typical basic and enhanced instruction sets. One of the primary distinctions between basic and enhanced is that the enhanced set used mnemonic functions a great deal to accomplish math and other data management functions. While it is possible to accomplish some of the same functionality with the basic set, it is more efficient to use these higher level instructions to perform the higher order functions.

Continuing with examples in the basic set, we look next at the latch function. Similar to the seal circuit noted above, the

Table 9.1 Instruction Set Examples

Basic	Enhanced	
RELAY	MOVE	REM—FM—BOT
TIMER	MOVE RIGHT 8	REM—FM—TOP
COUNTER	MOVE LEFT 8	SORT
LATCH	DP ADD	AND
ONESHOT	DP SUB	IOR
I/O REG	ADD X	EOR
REG I/O	SUB X	INV
BIN—BCD	MPY	MATRIX COMPARE
BCD—BIN	DVD	BIT SET
ADD	GREATER THAN	BIT CLEAR
SUB	TABLE—DEST	SHIFT RT
COMPARE	SRC—TABLE	SHIFT LT
MCR	MOVE TABLE	DO SUB
SKIP	ADD—TO—TOP	RETURN
		DO I/O

latch retains its logical state from the time power is removed from the system to the time it is restored. As noted above, the seal does not. The choice of using a seal or latch in a program design depends primarily on the need for this "retentive" feature. For example, certain safety precautions in a system design may dictate that a logic circuit *not* return to the "on" state after a power failure, but must be manually reset. (See Figure 9.2, along with the latch function description to follow.) When the START contact is momentarily closed, the LATCH (L) engages, and the real-world MTR becomes energized. When the STOP contact closes, the UNLATCH (U) activates, resetting the LATCH and disengaging the MTR. The function remains in this state until the start contact transitions again.

Other common basic functions include timers and counters. These software functions replace their electromechanical counterparts in many applications requiring counting and/or timing of discrete events. This could be the counting of parts on a conveyor or the time phased startup of a sequence of pump motors. In the programmable controller, counters and timers are configures in a similar way, described below and in Figure 9.3. In the counter, the count contact transition is sensed by the function and is registered as an increment (counters can be an up or down type, in this case we are assuming an up type) to the accumulated (ACCUM) count value shown in the lower rung. The preset (PRESET) value is established as a function of the programmer's requirement, and is placed as shown on the top rung. The counter function compares the preset and accumulated value of the counter and when

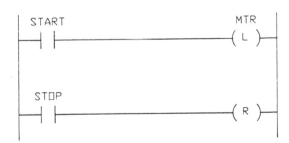

Figure 9.2 Diagram of latch function.

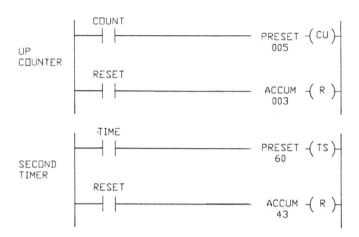

Figure 9.3   Diagram of timer and counter.

equal causes the output, in this case labeled FULL, to become
energized. Like all outputs in programmable controller logic, this
one can be used as a real-world output, or can be used as an internal "coil," fed back to other logic in another rung. If the accumulated count reaches the preset value, the output, labeled here as
FULL, will be energized. A momentary transition of the RESET
contact will reset the accumulated value of the counter to zero,
and turn off the output if it was on.

Timers operate in a similar fashion. When the START TIME
contact is closed, and remains closed, the value of the ACCUM
begins to increase by the time unit chosen. (Timers in programmable controllers are configured in seconds, tenth-seconds, and in
some cases, hundredth-seconds.) If the accumulated value reaches
the preset without the RESET contact closing, the output, in this
case labeled RUN, will be energized. RESET causes the output to
de-energize, and forces the ACCUM to zero.

The next group of programming functions generally included
in most programmable controller instruction sets are the data-in-
put/output/conversion functions. Examples are shown in Figure
9.4. As used here, these functions are referred to in mnemonics
are used with other relay ladders. Examples here are functions
that bring information in from Input references and move it to a

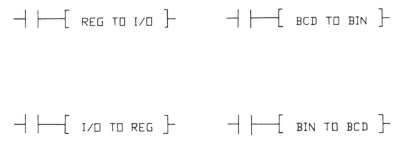

Figure 9.4 Diagram of data input, output, and conversion functions.

Register location (I/O to REG), and complementing ones that take information from Register references and move it to Output references (REG to I/O). As a variation on the same functional pair some instruction sets offer the ability to move information as described above, while at the same time converting it from Binary Coded Decimal to Binary, and the reverse. This is shown here as BCD to Bin, and Bin to BCD. These functions are useful to bring information in from external devices such as thumbwheel inputs, store it in register format, and later route it to an external device such as a digital display device.

Functions included in the basic instruction sets of some programmable controllers are for simple nonsigned integer math. These, like the preceding ones are generally of mnemonic format. Figure 9.5 shows a simple example of add and subtract functions. In this example, the contents of register references associated with A and B are added or subtracted as indicated, and the sum or difference is stored in the register reference associated with C. Again, as in most ladder logic, execution is performed upon "power flow," in this case as the normally open contact closes. These math functions are useful to compute the sums or differences of information such as the value of two timer or counter accumulators, or the values of two thumbwheel input devices.

It is at this level that most programmable controller manufacturers make the transition to their enhanced or extended instruction sets. In review, basic sets normally contain relays, latches, timers, counters, I/O to Register, and math functions. As we will see below, typical enhanced or extended instruction sets include functions such as extended data moves, table moves, list

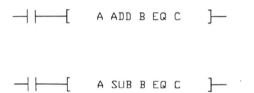

Figure 9.5 Diagram of basic mnemonic math functions.

managers, single- and double-precision four-function signed arithmetic, matrix manipulation, and subroutines execution. It is clear
from such a list that modern programmable controllers are capable
of doing much more than simple relay replacement functions. We
will examine typical enhanced functions below in logical groups,
looking in some cases at how such functions might be used. All of
the enhanced functions operate on a power flow basis, sometimes
referred to as a permissive contact basis. This is similar to the way
basic functions are utilized.

The first such group is the data move group. It normally
consists of functions that are designed to manipulate register data,
although some programmable controller designs allow use with input and output references as well, allowing a more efficient use of
memory in some cases. Included in the data move group generally
are functions for moving as single sixteen-bit reference from one
memory location to another, in this case from location A to location B (see Figure 9.6). The move function is useful for copying
individual data references from one source to a single or multiple

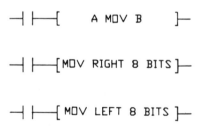

Figure 9.6 Diagram of mnemonic data move functions.

locations in the programmable controller memory, as might be done in certain closed loop control applications where a single feedback reference is used as an error reference for multiple loops. The next pair of functions usually included in a group like this are the Move Right 8-Bits and Move Left 8-Bits functions. These are also illustrated in Figure 9.6. This function allows eights bits of a 16-bit reference to be stripped away, without disturbing the eight bits that are left. These eight bits, whether left or right, can then be relocated to another memory location. This type of instruction seems to be most often used with ASCII-type data where each character used 8-bits or one-half of a sixteen-bit reference. This means that ASCII encoded information can be efficiently packed (two characters to a reference), into register memory, and then be serially spooled out, a character at a time, to a printer or CRT display, for example.

The next group to consider is the signed arithmetic group. Figure 9.7 shows examples in this group. These functions operate similarly to the Add and Subtract functions shown in the basic function set, but expand on them by including Multiply and Divide (MPY and DVD), on a signed (plus or minus) integer basis. In some of the more advanced instruction sets this group will also include the ability to operate with double-precision references. In this case that means a 32-bit binary reference, consisting of a one-bit sign and 31 bits of data. This offers an effective operating range of −2,147,483,648 to +2,147,483,647 in decimal. Single-precision arithmetic in this case means a 16-bit reference, with a one-bit sign and 15 bits of data. This predictably reduces the operating range, in our case to −32,768 to +32,767. And while it can be argued successfully that both are large number ranges, the single-precison range would be unsuitable for applications involving the manipulation of large amounts of data or data that can be used scaled, retaining the accuracy of many digits. Examples of applications requiring this are some high speed testing systems and certain process control applications. An optional member of the signed arithmetic group in some controllers is the Greater Than function. As the name suggests, this function tests two references to determine if one is greater than the other. If the conditions are true then "power" flows "through" the function ultimately to a coil that is used elsewhere in the ladder logic.

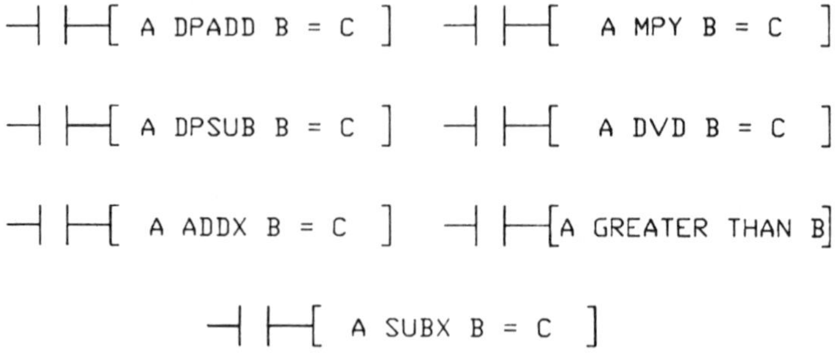

┤ ├─[A DPADD B = C]   ┤ ├─[A MPY B = C]

┤ ├─[A DPSUB B = C]   ┤ ├─[A DVD B = C]

┤ ├─[A ADDX B = C]   ┤ ├─[A GREATER THAN B]

┤ ├─[A SUBX B = C]

Figure 9.7 Diagram of mnemonic signed arithmetic functions.

The next group in most controllers is the Table group. This group is yet another example of the advanced data handling capabilities available in today's programmable controllers. The Table Functions allow the creation and rapid editing of tables of numeric data. Like all storage functions in the controller's memory, data here is stored in binary form and, therefore, can be encoded and decoded from and to any other form, such as decimal or ASCII. (If this is not clear, please review Chapter 3.)

Like many of the other enhanced or extended functions, the Table group employs the use of mnemonic references to build the functions. See Figure 9.8 for examples of the Table group. The

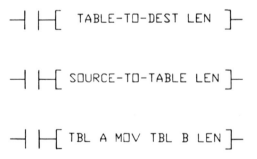

┤ ├─[TABLE-TO-DEST LEN]─

┤ ├─[SOURCE-TO-TABLE LEN]─

┤ ├─[TBL A MOV TBL B LEN]─

Figure 9.8 Diagram of mnemonic list functions.

functions, as they are created in most controller instruction sets, are used by defining a length (LEN) of the table of data required. A pointer function imbedded in the instruction keeps track of the table element it is currently accessing, and allows the instruction to act as a "data manager." Included in most Table groups are:

(1) The ability to take an individual piece of information selected by the pointer and send in to a certain destination memory reference (Table-to-Dest);

(2) The ability to do the reverse; that is take an individual piece of information from a given memory reference source and put it into a position selected by the pointer (SRC-to-Table); and

(3) The ability to move an entire table from one memory location reference to another.

All of these functions work in the now familiar power flow format, so we could describe a Table-to-Destination example as follows. With the contact permitting power flow to the function, the function would command the table element currently referred to by the pointer to be sent to the memory location destination specified by the reference. This normally occurs in one scan and then the pointer increments to the next element of the table. If, on the next scan the contact is again allowing power flow, the operation will be performed on this element. Table functions are most often used to structure a reference table of data used in a control process. This data is accessible rapidly and the results of the access combined with a formula execution cause a certain control function to occur or be modified.

The list group is the next function group we will consider. Figure 9.9 shows examples of this group. Lists are formatted and used in a manner similar to that of Table functions (above). The best way to describe the important differences between the two is to examine the way that they are designed to operate. The Table functions are formatted as static groups of information which could be accessible by selecting a single element anywhere in the group, regardless of position. The List function, however, is designed to allow dynamic movement of data through the structure, providing access by bringing a continuous stream of data from the top or bottom of the list. Hence the instructions described in the illustration as Add-to-Top, Remove-from-Bottom, and Re-

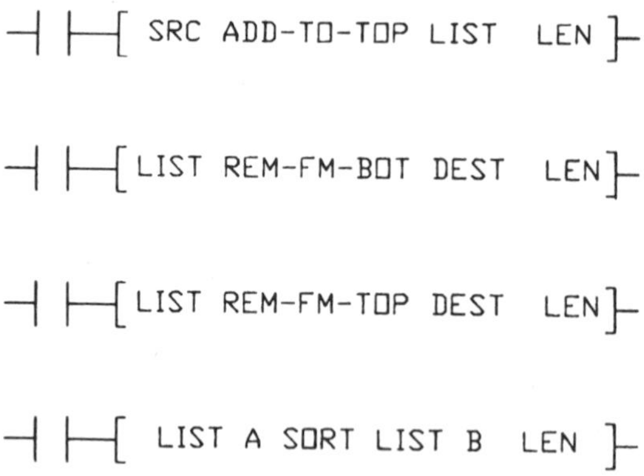

⊣ ⊢─[SRC ADD-TO-TOP LIST LEN]─

⊣ ⊢─[LIST REM-FM-BOT DEST LEN]─

⊣ ⊢─[LIST REM-FM-TOP DEST LEN]─

⊣ ⊢─[LIST A SORT LIST B LEN]─

Figure 9.9 Diagram of mnemonic list functions.

move-from-Top. The pointer in this case is used to "point" to the current top or bottom of the list, whichever is appropriate. Lists are used, as might be imagined, to gather and manipulate data that is collected sequentially. Using the Add and Remove functions, FIFO (first-in-first-out) and LIFO (last-in-first-out) executions can be accomplished on near real-time process data. In this way, trend data can be collected efficiently and analyzed quickly, allowing dynamic changes to be made to the process. Depending on the process design and needs, either the oldest or the newest data can be extracted from the list for analysis. In some controllers there is even a Sort function to allow some rudimentary statistical work to occur.

A comprehensive example is useful at this point, considering the ground we've covered so far. Figure 9.10 shows such an example, and work beyond this should be pursued with a text dedicated to ladder logic programming or a programmable controller manufacturer's course. The start pushbutton in the figure energizes the output coil labeled "operate." This coil continues to be energized until the stop pushbutton is pressed. The coil enables the second timer which has a preset of 30, which indicates that the coil on the timer (labeled "sample") will be energized after a duration of 30 seconds. In this case, the output from the timer is "connected" to the reset line of the timer, which means that the timer coil will be "high" for one scan and then be reset, in this

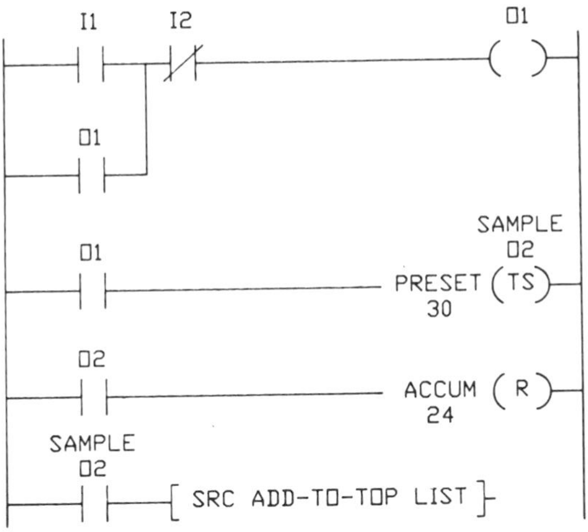

TIME BASED SAMPLE LOGIC

Figure 9.10 Diagram of comprehensive ladder logic example.

case, beginning the timing function over again since the enable contact is still energized. This sample command causes the Add-to-Top command in the next line to be executed every 30 seconds, causing the most recent data, in our case perhaps a temperature reference, to be added to the top of the list, while all older data is shifted down the list proportionally. The oldest piece of data ultimately "falls" off the bottom of the list, unless other provisions are made. This is not a concern in some cases, as long as the list is of sufficient length to collect a representative amount of data.

The next group usually included in the enhanced instruction set is the Matrix group. This very broad and powerful group allows the formation and manipulation of single or multiple binary bits of information. These functions can be utilized for as simple an application as tracking individual items on a conveyor system to performing complex diagnostic patterns on input and output information. Figure 9.11 illustrates matrix functions as they would be performed on multiple bits of binary information. The functions are labeled as AND, IOR, EOR, and MASK COMPARE, as examples of what might be typically included. AND is the operation that logically "ands" the two binary matrices referenced

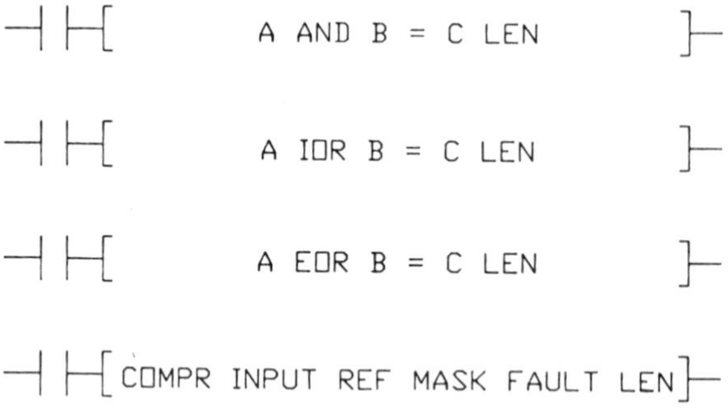

⊣├┤⟦   A AND B = C LEN   ├─

⊣├┤⟦   A IOR B = C LEN   ├─

⊣├┤⟦   A EOR B = C LEN   ├─

⊣├┤⟦COMPR INPUT REF MASK FAULT LEN⟧├─

Figure 9.11 Diagram of mnemonic matrix functions.

by A and B, and puts the result into matrix reference C. This means that each bit in each respective matrix is combined in this fashion. (For a review of Boolean Logic, refer to Chapter 2.) Similar functions are defined as IOR for Inclusive OR, and EOR for Exclusive OR. The last function references in the figure is the Compare with Mask function. Like all of the matrix functions, this one operates on two matrices. This one also uses a fault reference to contain any individual locations that do not compare, and a mask to "filter" out the miscompares that do not matter to the control system designer. This particular function, while not included in all programmable controller instruction sets, is a very powerful function for performing diagnostic routines while the controller is operating.

The next piece of the matrix group deals with individual bit manipulations. Figure 9.12 shows some examples of this group. This function includes the ability to set or clear, that is, change from a 0 state to a 1 state and back again, and to shift right or left a variable amount. These are extremely useful for applications involving extensive material handling of a large amount of units in the control system at a given time. Setting or clearing an individual bit in a matrix can simulate the presence or absence of a unit in the physical system, and the shift functions can be used to simulate the movement of a row of units in a conveyor system moving a variable distance.

The last group usually present in an enhanced instruction set is the control and/or subroutine group. This very advanced and

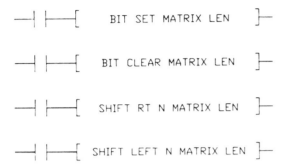

Figure 9.12 Diagram of mnemonic bit matrix functions.

powerful group of functions is designed to allow the programmer
to design variable execution paths for the ladder logic. They normally allow true subroutine development, including selective execution, iteration, and returns to the point of departure; and the
ability to selectively suspend and execute selected groups of I/O
points on the programmable controller's I/O system. This collection of functionality is normally used to provide very rapidly
executing program segments that are used infrequently. An
example of this would be a complex emergency shutdown sequence of a complicated and rapidly changing manufacturing process. Such applications require control logic to be able to interrupt itself, perform special logic and control functions rapidly,
and then return control to the normal ladder logic control program.

9.2 GRAPHICAL FUNCTION CHART PROGRAMMING

One of the newer forms of programming, at least in the United
States, is graphical function chart programming. This approach
is designed to easily and clearly describe, program, document,
and troubleshoot the control sequencing process. In Europe, this
programming method has enjoyed substantial development, and
is generally referred to as Graphcet programming. Graphcet is the
French standard for programming, and is used extensively there,
as well as in other parts of Europe. It is described as a superior

way to represent sequential logic, while relay ladder logic is normally best (after Boolean) for describing combinational logic. Since sequential processes dominate the majority of today's control processes, graphical function chart programming will almost certainly gain momentum as the world's standard for programmable controller programming methods. In fact, the International Electrotechnical Commission (IEC), along with the National Electrical Manufacturers Association (NEMA) are currently holding joint sessions to facilitate the development of this new standard.

Figure 9.13 illustrates an example of this future world standard. Graphical function chart programming allows the control design engineer to design the control process by simply describing it. Use of the language involves the employment of a new set of symbols and conventions, most of which are self-explanatory and include steps, transitions, connectives (also called directed links), and conditions. Steps are individual sequential symbols, represented here by numbered squares, which may contain mnemonics describing the function of the step. Transitions describe movement from one step to the next, and are described here by short horizontal lines. Connectives show the flow of control which is from top to bottom, unless otherwise indicated. Conditions are shown associated with transitions and may be written to the right. The example illustrates how easily a sequential drilling process can be described and programmed. The operation starts with a piece being loaded, then clamped, drilled, and removed, followed by a station rotation before the process begins again. Each square contains control commands that describe the discrete I/O and/or arithmetic operations that are programmed to occur. As can be seen from this simple example, this programming approach encourages a closer link between programmer and process designer, as well as offering a self-documenting program development process. Properly designed and implemented in the programmable controller, the approach offers immediate fault location and highlighting. This makes the approach ideal for the system integrator and system user. The graphical function chart approach is likely to become quite popular in the future. This presumed change is likely to form an overall approach which would include a slow metamorphosis from Relay Ladder Logic to a combination of

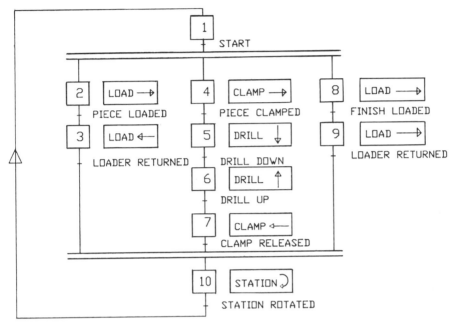

Figure 9.13   Diagram of graphical function chart programming.

Control Block Language, Statement List, and other higher order languages. More on this in Chapter 13.

9.3   BOOLEAN LOGIC PROGRAMMING

Boolean language programming is one of the more obscure languages insofar as it is used as a programmable controller language. This is perhaps curious in light of how well it relates to combinational and sequential logic routines, but understandable considering the more common and universal relay ladder logic evolving from electrical contacts and coils. A quick review of Chapter 2 will provide an overview of the functions utilized in Boolean Logic; including AND, OR, and NOT on the combinational side, and TIMER, COUNTER, and LATCH on the sequential side. These functions can be contrasted very closely to their relay ladder logic counterparts. Specifically this involves the use of AND functions as two contacts in series, OR functions as two contacts in parallel, with TIMERS, COUNTERS, and LATCHES utilized in a similar manner. An example of this is shown in Figure 9.14. Some industries use Boolean as a standard for control logic design, and certain European countries still prefer this form of programming. The programmable controller manufacturers in those countries sometimes offer Boolean programming as an alternative to ladder logic in some cases.

9.4 SPECIAL APPLICATION PROGRAMMING

As programmable controller manufacturers continue to develop new and differentiated product solutions, it is inevitable that special programming techniques related to unique application areas are developed as well. This has become even more common with the increased use of personal-computer-based programming devices. Examples of the application areas that are experiencing more special programming technique development are motion control and process control. The objective of any special application programming solution is to provide a man-machine interface that communicates to the programmer and/or operator in a language that he understands and relates to.

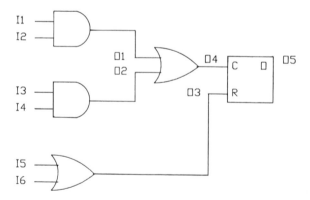

Figure 9.14 Diagram of boolean programming example.

In a motion control application, the programmer is interested in establishing the system and motion profile parameters in a straightforward manner. This includes, ideally, the ability to "converse" with the programming system in an English language customized for motion control jargon. In a programmable-controller-based motion control system, the task normally involves the integration of closed loop motion control with sequencing ladder logic. Well-designed application programming will allow the design of the motion control portion in "motion" language and the sequencing portion in ladder logic. Terms like acceleration, deceleration, end of limit, fly by move, and canned cycle are common to the engineer/programmer involved in motion control; they should be common to a special programming language as well. Figure 9.15 shows the screen of a programming device employing a motion control programming system. Motion control is common to a wide variety of flexible and dedicated machines.

Process control is perhaps more broad in nature, but is approached in much the same way. The man-machine interface "speaks" to the programmer or operator in terms that relate to process control, such as setpoint, process variable, output, and alarm. The system is designed to allow the average person working in the industry to both set up and operate the control process. This approach is common to both batch processes like plastic molding control, and continuous processes like paper mills and primary metal facilities.

SLIDE AXIS NO. 2

SET UP DATA:

```
+ END OF TRAVEL = +12 INCHES
- END OF TRAVEL = -2.5 INCHES
MAXIMUM VELOCITY = 600 INCHES/MINUTE
MAX ACCELERATION = 60 INCHES/MIN/SEC
SLOW/JOG VELOCITY = 6 INCHES/MINUTE
RESOLVER OFFSET = 0
REVERSAL COMPENSATION = 0
ANALOG LIMITS = 100%
```

Figure 9.15 Diagram of special application programming, motion control.

9.5 PROGRAMMING LANGUAGE ALTERNATIVES—A CONTRAST

To understand the full scope of the programming language alternatives described in this chapter, it is valuable to consider an example of a programming task, contrasting ladder logic, graphical function chart programming, and Boolean programming. The programming task is described below, and the alternative methods of solving the programming task are shown in Figure 9.16.

The programming task, admittedly simple, is to take a combination of input references including pushbuttons (PB1), rotary switches (SW2 and SW3), photocells (PC4), and limit switches (LS5), and produce an output (RUN). The control process calls for RUN to be energized when either LS5 is closed, or when PB1 and SW2 and SW3 are closed, but only if PC4 is not altered (that is, if PC4 in its normal state). In the case of relay ladder logic (Figure 9.16(a)), this becomes a simple series parallel arrangement of the contacts and coil. PB1, SW2, and SW3 are placed in series with LS5 in parallel with that combination. The entire network is then placed in series with a normally closed PC4, and the output coil RUN completes the task. Figure 9.16(b) shows the same task in graphical function block language. Almost trivial for this technique, the two states reflect the transition and energizing of RUN. In its fully documented state, function block 2 would contain the "logic" to determine the state of RUN. The third alternative, Boolean, is shown in Figure 9.16(c). Almost a direct translation of ladder logic, PB1, SW2, and SW3 are combined in an AND gate. This output is combined with LS5 in an OR gate. The output of the OR is logically ANDed with NOT PC4, with the result being the RUN signal.

While it is likely that each method has its strong and weak points, depending on the particular application and the skill and experience of the programmer, they all will have a place in programmable controller languages for some time to come. Even if graphical function block language emerges as a world standard, it is likely that the insides of function blocks will continue to be described in either ladder logic or Boolean format. It is also unlikely that any world standard would completely offset special application programming techniques since they are, by design, the most accurate for a given application.

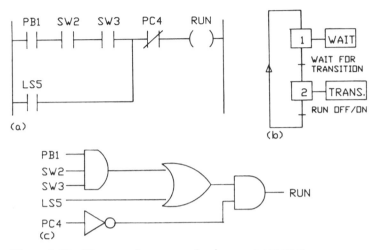

Figure 9.16 Diagram of programming language contrast.

9.6 TRAINING OPTIONS

While a text such as this provides an introductory overview of programming techniques, it is by no means intended as a substitute for formal programmable controller training. There are three basic ways that one can obtain such training. First, some firms offer self-paced text-based training designed to be taken at the leisure of the student. These are generally structured to include modules of training with review questions interspersed liberally in the text. This training offers the student perhaps the most flexible and inexpensive method to obtain introductory training. It has the disadvantage that no formal lab work is included, an important component of comprehensive programmable controller training. The second way to secure training is a "high tech" outgrowth of the first. It, like programmable controller technology itself, takes advantage of the common use and availability of personal computers. Some manufacturers of programmable controllers, as well as third party firms, are offering personal-computer-based training. Designed to be interactive in nature, this training can act as both instructor and lab partner. By asking questions about a module of text, the computer can assess the level of understanding and the retention, starting with simple review questions, and building up to more advanced problems. As a lab device it can simulate a real programming environment, while checking the student's responses automatically. The third way is perhaps the most common in use today, that is classroom based training provided by the programmable controller manufacturer, or a third party that specializes in training on a variety of controller brands. A typical course (or courses) will include, among other topics, the following:

Basic relay logic programming
Basic mnemonic programming
Extended or enhanced programming
Advanced application programming

Experience has shown that the opportunity to train with other individuals and interacting with human instructors provides the optimum method. And while the alternative methods are growing in popularity and use, it is unlikely that formal classroom training will decline. In fact, considering the proliferation of programmable controllers, it is more likely that the demands for all types of training will continue to accelerate for some time.

10
Installation and Maintenance

Today's programmable controller, properly installed, will provide a maximum of productivity with a minimum of maintenance.

This chapter will cover the straightforward but critical areas of installation and maintenance of the programmable controller. It is intended to serve only as an overview on the subject, with the best guide for use being the documentation provided by the programmable controller manufacturer. This chapter will include rack installation for the CPU and I/O racks involved, line power, grounding, and signal cable consideration. Troubleshooting techniques as well as basic repair situations will be examined, and finally, training options will be presented.

10.1 RACK INSTALLATION

Depending on the size of the programmable controller being considered, the installation of the racks or chassis can be a simple or very complex task. Since most controllers are of open or partially enclosed design, it is assumed that the proper cabinet is chosen for the particular application area. Many times this is a NEMA 12 type enclosure. It provides an environment in which the controller can operate without exposure to the grime outside the enclosure. Most racks can be mounted in either a panel type mounting arrangement, or a 19-inch rack mounting. This is not true of controllers of the very small variety, as they are normally a panel mount only design. Rack mounting is generally used where the equipment is installed in a control room where operators reside, or the environment is otherwise controlled. This type of installation is common where many instruments are mounted and used along with the programmable controller equipment. In either case, wire conductors must be routed in the rack; this is done with commercially available conduit that provides a means to bring in and out as many as 100 to 200 individual conductors from input and output points to real-world sensors and actuators. It is important that this phase of the design and installation be handled with care as it will dramatically affect the ability to maintain the system later. Good shop practices will provide carefully labeled and bundled conductors, routed through properly sized conduit systems. In addition to allowing any required system maintenance, this early care will make any system additions or modifications much easier. Figure 10.1 illustrates a typical chassis installation.

10.2 LINE POWER AND GROUNDING

As we saw in Chapter 4, proper power to the programmable controller is critical. Today's systems are available in a wide variety of electrical configurations. Virtually all are designed for use in single phase power systems, and most are now beginning to be offered with the optional ability to operate in a DC supply environment. AC designs are offered in either single voltage supplies, such as 115 or 230 volts AC; while some can be configured as either through a selection made on the power supply. Figure 10.2 shows a typical power supply connection scheme. Proper grounding of the power supply connection is required for a safe installation. Some programmable controller designs have individual grounding connections from rack to faceplates and other system components, so care must be taken to follow good electrical practice in system grounding during electrical installation. In certain

Figure 10.1 Photo of a programmable controller system installed in a cabinet. (Courtesy of General Electric.)

177

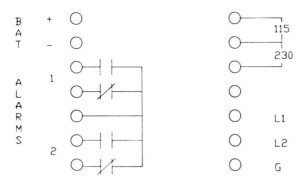

Figure 10.2 Diagram of a typical power supply connection.

applications, a 24- or 120volt DC power supply is required. This is common for installations that are made where no AC power is available, such as remote electrical generation stations. It is also found where AC power is unreliable and where loss of control is considered an unacceptable situation.

10.3 SIGNAL CABLE CONNECTIONS

Chapter 7 touched on the importance of having good signal cable connections from chassis to chassis. The chain must be complete and have high integrity to provide the communication path for control signals to pass over. Depending on the specific design of the programmable control system, loss of communication to the I/O system will cause a critical failure, stopping the CPU scan. In other systems, configurations can be accomplished that allow the unaffected portions of the programmable controller system to continue to operate.

As we saw earlier, communication between chassis can either be parallel or serial. Parallel communication uses multiple conductors to pass all bits of a byte or word of data simultaneously. Serial communication provides a method for single bits of a byte or word to be transmitted sequentially. The difference this has on cable design and selection is significant. Parallel cables are made of as many as 16 or more pairs of twisted individual conductors,

protected by a shielded external jacket. Multiple pin connectors
are connected to one or both ends of the cable. The "D" type are
popular, providing 25 or 39 pins in a standard configuration.
Earlier designs used a round type positive contact threaded design
and certain manufacturers still use this type of connector, although most have found the "D" type satisfactory. This cable is
normally made available by the programmable controller manufacturer, and while it may seem expensive, the assurance of having
a well designed and thoroughly tested cable is worth almost any
price compared to tracking down an intermittent problem later
during a critical production period. The programmable controller
manufacturer will provide parallel cables designed to operate with
its equipment in various lengths, from 2 feet up to as much as 500
feet. Serial cables on the other hand are generally easier to use
and less expensive to purchase or build. Manufacturers may or
may not provide these as an item to purchase, and may just refer
the purchaser to a third party source. Serial signal cable are normally a single or dual twisted pair of conductors, and are often
just connected directly to the I/O driver or receiver with screw
terminals. The only major reason to use parallel over serial is the
greater communication speed available with parallel, which can be
a factor of 10 to 1, or more. Figure 10.3 shows a typical cable
connection scheme.

10.4 TROUBLESHOOTING AND REPAIR

Even the best of today's well-designed and manufactured programmable controllers require occasional preventative maintenance and
repair. This section looks at some of the tools provided by the
manufacturer and techniques for general maintenance.

Most of the medium- and large-sized programmable controller
systems available today are designed to be maintained by individuals with a wide variety of skills, without the benefit of in-depth
formal training of this piece of equipment. This is accomplished
in the design by providing individual modules of functionality
installed in a chassis serviced from the front (all module types including power supplies). Front access is critical to proper maintenance. This allows easy inspection and replacement of the suspected bad module. Module health is determined by inspecting

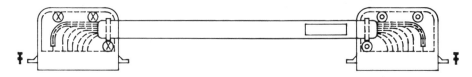

Figure 10.3 Diagram of a typical parallel cable connection scheme. (Courtesy of General Electric.)

the suspected bad module. Module health is determined by inspecting the LED indicators normally provided on the front of each module. Typical indicators will be on or off depending on the design and individual condition of the module in question. Various CPU and I/O modules will have indicators showing I/O control communications status, memory integrity, power supply tolerance check, scan integrity, and others. On future controller designs, and even today on a few systems, it is likely that English language messages will be displayed on the controller advising the user or maintenance personnel that a particular failure has occurred and recommended actions to take.

The modular design and diagnostic indicators are, of course, important, but would be quite useless without well designed documentation provided by the manufacturer for the programmable controller system in question. Proper documentation will have sections dedicated to each major subsystem including CPU, I/O, and programming device. Each should explain in depth the step-by-step inspection of the system. All possible combinations of failure mode should be listed, along with suggested actions for repair. This will most often involve only the substitution of a replacement board for the suspected failed unit. The user is urged to purchase a set of spare modules for the system in question as recommended by the manufacturer. This is normally, at a minimum, a single replacement module for each CPU and programming device serviceable module, and spare I/O modules equal to 10% of the number in the system.

While the design of system and documentation provide an easy vehicle for system maintenance once trouble is detected, it is always better to perform preventative maintenance where possible. Some systems available today provide levels of predicted

Figure 10.4 Photo of a typical formal classroom scene in the use and maintenance of programmable controller systems. (Courtesy of Texas Instruments Inc.)

failure detection that allow excellent preventive maintenance. This can range from automatic signalling of low battery conditions on the memory modules to predicting an output circuit failure before the circuit is energized. This particular phenomenon can be extremely valuable in performing predictive maintenance on critical control systems. Examples of this would include emergency circuits for rapid shutdown of a machine sequence or plant process. Outputs that fail prior to their need may not be detected in time to provide this critical service.

The documented ladder logic diagram is a very useful tool for troubleshooting, especially on a complex system. It allows maintenance personnel to "walk through" the I/O segment in question, forcing transitions in specific inputs andoutputs, while watching the system's reaction. Many times this can ferret out a

tough system problem, and may help isolate problems external to
the controller such as an open sensor wire.

The major manufacturers of programmable controller systems maintain 24-hour telephone service numbers for users requiring emergency service. Their highly trained personnel can normally walk the user through his problem and determine what action he should take for repair.

10.5 TRAINING OPTIONS

As with programming skill development, training is available for
general maintenance and repair of the programmable controller.
Most common is formal classroom training which provides an
effective combination of class study and lab exercises that simulate problems for the student to detect and correct. Figure 10.4
illustrates a typical formal classroom setting. Classroom training
can be done at the programmable controller manufacturer's facility, or, in some cases, can be more efficiently done at the user's
location. Supplementing this formal method are programmed
instruction books, videotapes, and interactive computer-based
instruction. This latter option is just now becoming available,
but holds great promise for effective training, easily distributed
and updated.

11
Applications

*"We are all continually faced with a series of great oppor-
tunities brilliantly disguised as insoluble problems."*

John W. Gardner

Applying programmable controllers is as critical to the user, as is
solving the problems that come with designing and operating
efficient production processes. In fact, it is through its innovative
applications that the programmable controller has excelled and
attracted the large following it enjoys today. Continued development in the application arena supported the evolution of certain
features and capabilities that are common today. In this chapter
we will examine the current scope of programmable controller
applications, briefly look at the steps involved in implementing a
system for a typical application, and consider some logical group-
ings of applications common today, including basic, industry
specific, and generic.

11.1 WHAT A PROGRAMMABLE CONTROLLER CAN DO

In Chapter 1 we saw a partial listing of programmable controller
applications, and while it was anything but complete, it did give
a good overall look at the breadth and variety of use that program-

mable controllers enjoy. These systems, with such a simple and low technology heritage, have come to be capable of performing tasks that were previously considered impossible to accomplish, or at least very difficult and at great expense. We will look at a survey of applications by major industry groups, and while some of the application areas are common to many of the groups, it provides a structured method to categorize the wide variety of applications. The structure will consist of two major groups—discrete part manufacturing and process—and several subgroups in each major group. Table 11.1 provides a summary of the structure.

Discrete Part Manufacturing Industries

Transportation. This industry group consists of automotive, aircraft, aerospace, railroad, and ship segments, as well as a number of smaller segments. Programmable controllers are used extensively in many of these segments, especially the automotive segment. Both the vehicle final assembly process and the various subassembly and part manufacturing processes make wide use of programmable controllers. Figure 11.1 shows an engine assembly process that uses programmable controllers for sequencing, data concentration, and communication. The largest part of these are used for sequence and motion control applications, combined with a growing overlay of data management and communications applications. Almost all of these are driven by metal cutting, forming, and assembly needs, although a growing trend is in the plastic molding and forming area, as plastics and other engineered materials are replacing metal in some areas. The aircraft and aerospace segments share the use of programmable controllers for metal cutting and forming, but are lighter on assembly since many of the assembly processes are custom oriented, and hence cannot justify large capital outlays in hard automation. The need for factory floor data collection and analysis is very intense here, and programmable controllers play a major role in front line data collection and concentration. Data is collected to document process integrity involving exotic and expensive metals, and for processes critical to design integrity or requiring regulatory agency documentation. The railroad and ship segments have less intense needs for programmable controllers, but one innovative use in the railroad industry involves the use of programmable controllers as

Table 11.1 Industry versus Application Intensity

Industry	Application Segment				
	Sequence	Motion	Process	Data	Communication
Transportation	H[a]	H	L[c]	H	H
Fabricated Metal	H	M[b]	L	M	M
Nonelectrical Machinery	H	H	M	H	H
Electrical & Electronics	H	H	M	H	M
Food	M	L	H	H	M
Chemical	H	M	H	H	M
Primary Metals	M	H	H	M	M
Paper	H	M	H	M	M
Electric utility	H	L	M	M	M

[a]H = High
[b]M = Medium
[c]L = Low

trackside signal and switch control units, operating remotely and communicating with a central dispatching station over phone or microwave links.

Fabricated Metal. The fabricated metal segment consists of a wide variety of sub-segments and processes, and as the name suggests, most involve the use of programmable controllers for metal forming and, to a lesser extent, metal cutting. The sub-segments range from manufacturing metal doors and windows to producing firearms. Controllers are most often found in sequencing and motion control applications, and much use is made to accomplish sophisticated material handling applications.

Nonelectrical Machinery. This segment was designed to contain a very loose arrangement of machinery manufacturers, ranging from the general to the very specific. It includes both farm and construction equipment manufacturers, internal combustion en-

Figure 11.1 Photo of engine assembly process. (Courtesy of Gould, Inc.)

gine manufacturers, manufacturers of machine tools, and an important category called special machinery manufacturers, which includes companies that produce special purpose machinery such as tobacco, shoe, elevator, and plastic molding. As you might gather, such a variety of manufacturers use programmable controllers in very large numbers. The way in which they are used is relatively straightforward, and includes a large measure of sequence control and motion control. Larger facilities are also taking advantage of the communication and data management capabilities of the more sophisticated controllers. Elevators use the motion control and communication capabilities. Packaging machinery many times can use the smallest of today's programmable controllers as their sometimes limited application needs can be served by a basic featured, limited I/O controller. On the other hand, some packaging applications require the fastest and most powerful programmable controller features.

Electrical and Electronic.   This segment is made up of manufacturers ranging from those that produce televisions and radios to those that make batteries. Machinery that is used in these manufacturing processes includes some batch processing, as well as a measure of sequential control (material handling), and motion control (component insertion). Other fertile areas include automatic assembly, automatic storage and retrieval systems, and testing systems. Again, as in the other segments we have examined so far, larger facilities are using local area networks to take advantage of the communication and data management features of the newer controllers. Also as with other segments, we find many, many stand-alone machine applications using small- to medium-sized controllers, including queueing conveyor systems and specially designed packaging machines.

Process Industries

Food.   The food processing industry is one that has a wide variety of programmable controller applications, ranging from the batch processing of fluid milk to the batch processing of beer, all the way to the packaging of both products. Material handling applications are common, and include both bulk (the controlled movement of a continuous stream of product), and unit (the controlled movement of individual units of product). PID closed loop control is common with the need to regulate flow and level of foods in process, as well as the temperature regulation of certain cooking and sterilizing processes. Government regulation of many food-related processes calls for meticulous documentation and data gathering of process parameters to guarantee proper food quality. All of these application needs call on the natural strengths of programmable controllers, and use a broad sample of control schemes. Figure 11.2 shows a cookie baking application that uses programmable controllers for a variety of material handling, and process regulation and control tasks.

Figure 11.2 Photo of bakery process. (Courtesy of General Electric.)

Chemical. The chemical industry consists of a number of related segments, including pharmaceuticals, paints, and engineered materials, such as plastics. This group relies on the programmable controller not for primary continuous process control needs, but is used in conjunction with the process computer. While the process computer has command of the many process loops involved in a continuous process, the programmable controller, connected through a communication interface, operates the various pump motors and supply valves involved in the process. In smaller facilities that produce products utilizing more batch processing, programmable controllers are sometimes used as the primary control devices. A, perhaps surprisingly, large application area in the chemical segment is the packaging area. Programmable controllers, properly applied, make excellent control devices for high speed packaging processes.

Primary Metals. Primary metals are aluminum, steel, etc. The production of these metals relies on programmable controllers for many and various needs. Most of these relate to the sequencing and movement of the metal, in its liquid, semisolid, and finished states. Continuous casting processes, hot and cold rolling processes, and coding and in-line storage of finished materials all use sequence and motion control heavily. Integration with variable speed drives is common. Also common are uses of controllers for control of slitters and coilers. In some facilities, programmable controllers are also used for batch mixing and weighing of materials at the front of the process. This involves PID control and regulation of weights and flows, as well as the integration of bulk conveyor control.

Paper. Paper industry applications include both the original creation from wood fiber, and the conversion to finished goods such as corrugated boxes and stationary grade paper. Programmable controllers are used here in ways similar to those in the primary metals segment. This should come as no surprise since both are web processes. The "wet" end of the process is now experiencing more applications of programmable controllers for loop process control. Figure 11.3 illustrates the chip handling process at a paper facility, which is controlled by programmable controllers. The coordination of multiple variable speed drives on typical a paper machine is accomplished by a combination of programmable controllers and special purpose controls. The converting process, normally asynchronous from the primary process, involves a number of cutting, folding, and assembly processes using motion and sequence control.

Electric Utilities. The electric utility industry, like many other process industries, uses a process computer for the actual control of the generation process, but uses programmable controllers to perform ancillary functions. These include the bulk material handling of coal into the generation facility, the removal of waste ashes from the process, and the control of the scrubbers used to clean the by-product process air. These are connected to the main process computer by a communication link.

Figure 11.3 Photo of paper industry application. (Courtesy of General Electric.)

11.2 IMPLEMENTING A PROGRAMMABLE CONTROLLER SYSTEM

This section will describe the steps in the process of integrating the programmable controller and the control problem. The implementation phase of using programmable controllers is, by definition, the most important. It is through the successful implementation of the programmable controller that it is best allowed to solve the problems at hand. In addition, the proper implementation will also prepare the system to accommodate control process needs in the future.

Table 11.2 lists the steps involved in implementing a programmable controller system. Below, we will describe each briefly, and show how each contributes to the implementation.

Table 11.2 Steps Required to Implement a Programmable Controller System

1. Describe the control process.
2. Define the inputs and outputs.
3. Estimate memory requirements.
4. Determine man-machine interface.
5. Determine instruction set required.
6. Estimate response time required.
7. Document the program.
8. Write the program.
9. Test and debug the program.

1. Describe the Control Process. The very first operation required is to describe the control process in a comprehensive and detailed way. The end product of this exercise will be a functional description of the control solution as it relates to the control problem. It includes the diagrams and text that describes the relation of the programmable controller CPU, its Input and Output, and the man-machine interface to the controlled process. It also includes any other intelligent devices in the system, including the basic connection to a host computer. The text going along with this segment should describe the process and all possible combinations. Figure 11.4 illustrates part of a typical process description.

2. Define the Inputs and Outputs. The second step involves the detailed listing of the real-world inputs and outputs of the programmable controller. Analog and discrete I/O needs are described and listed separately, and an individual listing group should be made for each field signal level, i.e., 115 volts AC, and 24 volts DC. Included on the listing should be slot address, circuit number, I/O point address, and circuit name. An accurately done job here will make the documentation process easier. In addition, logical grouping of the I/O points will make programming and use of the system much easier as well.

3. Estimate Memory Requirements. As we saw in the chapter on the CPU, the programmable controller uses memory to store the control logic program. It also uses memory to store

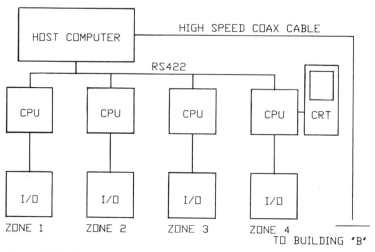

Figure 11.4 Diagram showing a typical process functional description.

variable parameters, such as timer and counter accumulations and
I/O status tables. Depending on the particular system used, memory sizing will require different approaches. As a rule of thumb,
simple relay replacement applications utilize one I/O point for one
control relay equivalent. Assume also that the average application
(if there is such a thing) uses inputs and outputs in the ratio of
6:4, and that one control relay equals eight contact references.
Since most 16-bit word systems use one word of memory for each
contact and reference combination, each relay replaced requires an
average of eight words of memory in the programmable controller'
logic memory. This suggests that a programmable controller system designed to replace a panel of 100 relays would require an
average of 60 inputs, 40 outputs, and 800 sixteen-bit words of
memory. Figure 11.5 shows a typical logic network and the memory used. Variable memory storage must be estimated separately,
and will range from the minimum normally specified by programmable controller manufacturers for simple relay intensive applications to very large requirements (say 8,000 to 16,000 sixteen-bit
registers) for very data intensive applications such as automatic
testing, and automatic storage and retrieval systems.

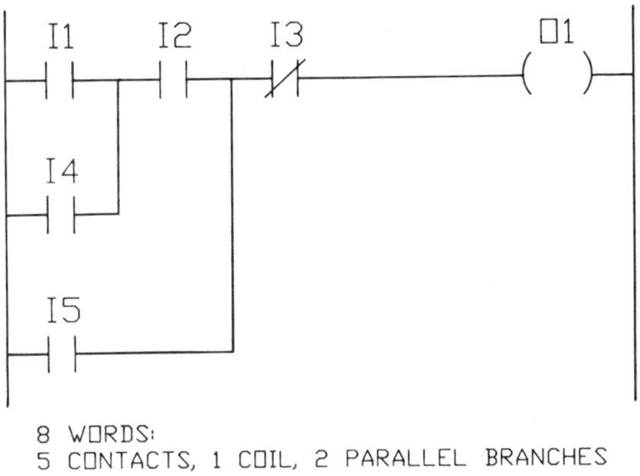

```
8 WORDS:
5 CONTACTS, 1 COIL, 2 PARALLEL BRANCHES
```

Figure 11.5 Ladder diagram with typical memory usage.

4. Determine Man-Machine Requirements. Over and above the requirement above would be any I/O and memory intended for an operator interface panel. If this is a simple requirement, it may be handled with a few real-world I/O for pushbuttons and indicator lights. These devices would be mounted in a panel or cabinet for use by the operator and would be wired individually to discrete (and, in some cases, analog) I/O modules in the programmable controller system. For a more sophisticated CRT based interface, a data communication interface will normally be required. Also important is the determination of the functional requirements for the CRT system. For example, if the majority of the interface requirements are to be performed by the CRT, that system must have capacity in communication throughput, and screen quantity and complexity to support those needs.

5. Determine Instruction Set Required. The application complexity also determines the choice of an instruction set. For small stand-alone machines, the basic instruction set provided by the manufacturer will probably suffice. However, for larger and more complex machines and processes, the enhanced instruction set is always a cost-effective investment. Considering the trend demonstrated by programmable controller manufacturers so far, it is likely that more features will continue to evolve in the form of better instruction sets. These instruction sets will be found, more and more, in smaller, less expensive controllers.

6. Estimate Response Time Required. For many applications the response time consideration is not a major issue since
most programmable controllers available today are capable of
handling a large number of I/O and logic solutions with sufficient
speed. For some applications, however, the speed of the controller is an essential consideration. Speed, in terms of solving problems, is measured as response time. Response time is the total
time required to convert the appropriate input to a given output,
including the following components: input filter time delay, I/O
service time delay, logic solution time, and output filter time delay. All of these components must be considered for time-critical
circuits. To estimate normal expected throughput, the scan time
can be doubled (or tripled for margin) to accommodate a worst
case response time. Improvements in response time can be made
through examination of the particular system being used. For
example, the input filters, designed to provide switch "debounce,"
can be circumvented in some systems. This normally contributes
greatly to improving throughput. Next comes the opportunity to
make the logic used more efficient through otpimization of ladder
functions, and the increased use of enhanced functions where possible. This improves the logic solution time as the enhanced functions generally execute more rapidly than do the basic. Output
filters are generally not a major contributor to system throughput,
but can be examined if necessary. All of these actions taken together can improve throughput by 30 to 50%. In some rare systems, hardwired interrupt inputs are provided allowing the input
process to be reduced to microseconds instead of milliseconds.

7. Document the Program. While it may seem odd to show
the documentation of the program prior to actually writing it, it
is actually quite pragmatic. The program should be documented
in terms of its relationship to the controlled process, and the operation and maintenance of that process. This will naturally include
diagrams, tables, and text, intended to provide the user with an
easy to use road map of the controlled process. A flow diagram,
timing diagrams for critical functions, annotated ladder diagrams
(generated later), and data flow diagrams give a detailed look at
the process. I/O definitions, wire lists, along with truth tables illustrate how the physical system comes together. And finally,
complementing text provides a common thread throughout for

the documentation of the system. The value of structuring the process should not be underestimated. A properly and thoughtfully designed structure becomes the basis for a better overall solution.

8. Write the Program. Now is the logical time to actually begin writing the program. With the proper effort applied in the steps identified so far, the program writing task is indeed minimized. The process flow outline can be clothed with appropriate ladder logic programming. From a size and speed efficiency standpoint, enhanced instruction functions should be used wherever possible, except where doing so would make the programming needlessly complex to maintain later. Proper annotation and text comments go a long way to avoiding any confusion. Subroutines, available in some controllers, can be used to structure the programming effort, and to make execution more rapid. Special considerations must be made in the areas of power-up initialization, power-down impact, and safety. It may be important to provide logic to reset certain functions during a power-up cycle, and to predict a stable power-down routine. Safety circuits should be hardwired, in spite of the high reliability of the programmable controller. Status tables can be modified unintentionally, or remotely, causing an unsafe condition.

9. Testing and Debugging the Program. The most well-designed program will still experience some flaws, some perhaps unpredictable. In light of this, it is important that the user or system integrator go through a series of steps prior to actually starting the machine or process. This includes testing the logic off-line in small sections to ensure that it responds predictably. Output wiring can be tested by forcing outputs on and off individually to ensure that field devices are wired properly. And finally, initialization routines and safety circuits must be thoroughly checked.

11.3 BASIC APPLICATIONS

In the program development stage of implementation, it is often valuable to be able to take advantage of basic application programming already completed and tested for the programmable controllers being used. Basic programs such as this are available many times from the programmable controller manufacturer in the form of an application manual or guide. These are not complete programs, but program segments designed to solve a particular but predictable problem. They are generally used by integrating them with custom programs that are developed for the application problem being considered. Included in this category of basic application are time of day clock, I/O data entry and display, averaging data samples, data table lookup with linear interpolation, scaling analog input data, alarm detection and annunciation, setpoint ramping, and shift register step sequencer. Utilization of these predesigned program segments can often save time, both in program design and later debugging. Figures 11.6 and 11.7 show two examples of basic applications; a time of day clock and a time-based step sequencer.

11.4 INDUSTRY SPECIFIC PACKAGES

As programmable controller applications have evolved over time, certain ones have clustered around combinations of hardware and

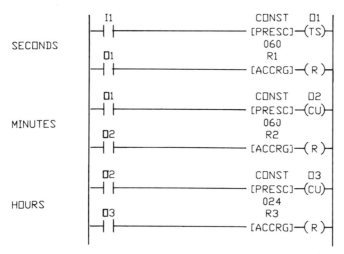

Figure 11.6 Diagram of time of day basic application.

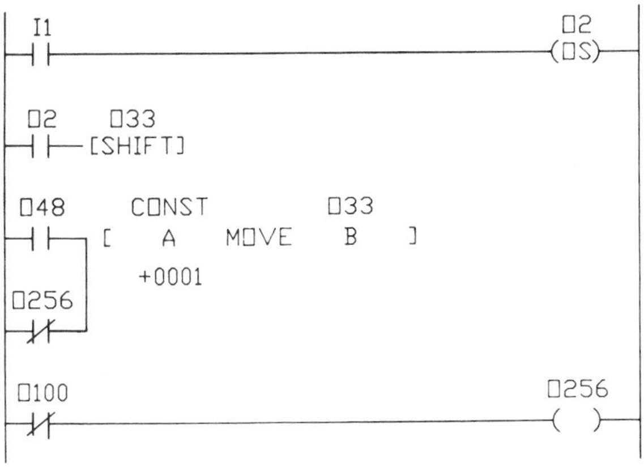

```
 I1                                          □2
─┤ ├──────────────────────────────────────(□S)───

 □2    □33
─┤ ├──[SHIFT]

 □48      CONST        □33
─┤ ├──┐  [   A   MOVE   B    ]
      │      +0001
□256  │
─┤/├──┘

□100                                        □256
─┤/├──────────────────────────────────────( )───
```

Figure 11.7 Diagram of time based step sequencer basic application.

software that were differentiable in the marketplace, aimed at specific industry segments. In some circles, these package combinations came to be known as "Green Packages," relating to the fact that the value of the package was in the *solution*, not in the specifics of the hardware or software which became somewhat transparent in the transformation. This evolution of transparency is a signal that the programmable controller industry is experiencing a phase of maturity, a blending of "customized" application solutions with an increasing number of "Green Packages."

Perhaps the best way to illustrate the concept is to categorize some of the package possibilities into application groups. We will consider the groups listed below.

> Motion control
> Batch process control
> Data management
> Communications
> Diagnostics

It might be noted that sequence control was not included on the list above, and while it follows logically that it should, this segment is almost always designed in a custom way. In fact, on all but the most simple applications, sequencing is the mechanism that ties together the logic for all of the so-called higher order application segments.

The general characteristics of these packages are that they (1) combine hardware and software together in a way that solves a spedific application or industry segment problem, (2) use ladder logic that represents 50 to 75% of the required solution, (3) require limited modifications, and (4) may include English language "front ends" allowing interaction with the system by those with limited training.

Motion Control.   This involves the use of appropriate CPU and motion control I/O, along with motion profile generation software to solve single or multiple axes control problems.

Batch Process Control.   This combination allows analog and digital I/O to work along with PID software to regulate batch type processes.

Data Management.   This is a broad solution, involving the collection and manipulation of data on the factory floor. The result is used generically for alarming, trending, and simulations, etc.

Communications.   This is simply an intelligent front-end that allows a user with simple skills to configure and use programmable controller communications. The set up of read and write sessions, along with other basic tasks are included.

Diagnostics.   Machine and process diagnostics have long been the quest of every process designer and shop foreman. With today's sophisticated controller, hardware and software can be designed to detect, analyze, and report faults, pinpointed to the lowest fault point.

The proof, of course, is in the actual execution of solutions, and while many of these combinations are just coming into practical use, there are many uses providing fertile ground for package solution. In the following sections, application possibilities are listed for motion control and batch process control.

Motion Control

Nonelectrical Machinery Industry
Grinder
Indexing table
Cartesian robot

Drilling machine
Screw machine
Transfer machine
Punching/Shearing machine
Bending/Forming machines
Automatic assembly machines
Broaches
Presses

Electrical and Electronic Industry
Wire coil winders
Component insertion
Testers
Assembly

Rubber and Plastic Industry
Treadline control
Film line
Tire building
Calendar control

Paper Industry
Converting machinery
Winders
Paper machine

Primary Metals Industry
Continuous caster
Hot strip mill
Cold mill
Coiler

Batch Process Control

General
Configuration package
Trend analysis
English language batching
Symbol library
Batch weigh
Alarm annunciation and logging

Rubber and Plastics Industry
Plastic injection molding
Extrusion
Calendar control
Tire press

Food Industry
Extrusion
Retort control

Paper Industry
Slitter coater, coater kitchen
Digester
Tie line control
Lime kiln control
Stock bleaching
Washing
Calendar control
Boiler auxiliary control

Electric Utility Industry
Tie Line Control
Substation/Scada control
Small hydro station control
Precipitator
Scrubber
Baghouse
Boiler auxiliary control

12
Communications with Programmable Controllers

Often hailed as revolutionary, industrial communication for programmable controllers is actually evolutionary, taking advantage of emerging means of communications to produce the "missing link" in factory automation.

The reality of communications between programmable controllers and other intelligent devices brings the long sought promise of productivity into crisp focus. This part of today's factory automation arena is characterized by a diverse array of terms including networks, local area networks, data highways. Regardless of what terms are used, communications are probably best defined in terms of the services they provide and the applications they make possible. In this chapter we will examine some of these services and applications, along with current standards and trends in industrial communications for programmable controllers.

12.1 INTRODUCTION

The concept of communications is not new in its application to office automation, and certain process control computer systems. For many years, office computer systems have shared large database files among multiple users. Computer manufacturers nurtured the development of proprietary networking methods which many times required standardization by the user on a particular brand of computer. In recent years, broad standards have begun to evolve in office networking, allowing many manufacturers and users to work toward a common framework in the communication arena.

It is only in the past three to five years that basic communications between programmable controllers and other factory floor devices has become popular. And, like the office evolution path, factory communications is currently changing from an environment that has many competing proprietary communication standards to one that has a common framework for manufacturers and users. Different levels of communication applications are being developed to handle the varying control and information requirements on the factory floor. For example, relatively low speed baseband networks are currently being applied to the discrete control level for the communication of actual I/O status from one controller to another, while higher speed baseband coax systems are being applied at the manufacturing "cell" or shop floor level, coordinating the activities of five to fifteen intelligent devices including, but not limited to programmable controllers. These "cells" can be connected together by a broadband coax grid, which encompasses the entire factory floor, providing essential connections to the plant's host computer system(s). In the most elaborate systems, these plant levels can be connected through wide-area communication systems at the "corporate" level. Figure 12.1 illustrates this industrial communications hierarchy.

Our focus for this chapter is on communications between programmable controllers, but it is important to note that the concepts presented here apply to many other intelligent devices such as computer numerical controls and robots. Today's programmable controller uses a microprocessor-based interface device to provide a reliable method to convert higher speed coax-based signals to formats and levels used at the backplane of the programmable controller. This interface can reside external to the device, or can be imbedded in the controller, which provides the preferred combination. The external, stand-alone interface can be used to connect the odd intelligent device to the network, allow-

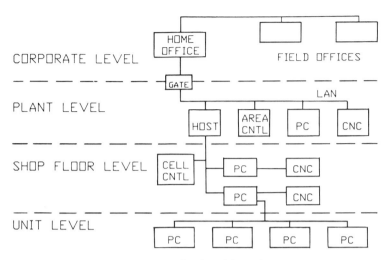

Figure 12.1 Industrial communications hierarchy.

ing incorporation of many devices in a single network. Industrial networking, or local area network (LAN) standards, such as those supported by the IEEE 802 structure, are providing a structure for all manufacturers to connect to. This evolution promises to bring about new affordable levels of productivity to manufacturing facilities both large and small.

12.2 CONCEPTS AND APPLICATIONS

As noted earlier, communication networks are best described by the services they provide, and the applications (solutions) they bring. It is appropriate then to examine some of the characteristics and applications at each of the levels: plant, cell or shop floor, and unit.

First, the plant level. Programmable controllers directly connected to this level of network will be rare. If present, they will normally be of a supervisory nature, perhaps of a noncritical application such as energy management, which also benefits directly from close access to the plant host computer, as it can then relate all areas of a plant for overall facility energy management. Most of the devices connected at this level will be computers. Some of these computers will be optimized with software to allow them to do area or cell control. Others will be dedicated primarily to batch office data processing, with data about production status and yield being updated from the factory floor occasionally. This is the factory "backbone" local area network, a broadband coax, outfitted with hardware and software to allow high speed, multi-channel communications. High speed here should not be mistaken for high throughput, since the multichannel broadband system will have a substantial amount of overhead, and will have to deal with some throughput limitations in the interface devices. Although time sensitivity and response is of some importance here, it is the least important of the three levels, since no actual machine or process control occurs here. The primary users at this level are humans, normally factory management, and human interface is forgiving on time response, where machinery is not.

The shop floor, or cell level, network is likely to be home for a larger number of programmable controllers than the plant level, and time sensitivity is more critical here as well. This level of control will use a local area network to orchestrate the various PCs, NCs, robots, vision systems, and other general purpose or dedicated devices. The orchestration involves insuring that all systems are operating properly, have the appropriate application programs loaded, have materials to operate on, and are maintained at the right intervals. This network will also likely be a coax type, but will probably be baseband in design, and dedicated to serving the devices only in the cell. The users of this level, therefore, are primarily nonhuman, and therefore less forgiving in terms of response time for production tasks.

The unit level is the lowest of the three, but the most critical and least forgiving in terms of time response. Programmable controllers are very common at this level, normally more than any other device. It is at this level that actual I/O status and control signals are sent from one programmable controller to another. This might be the cooperation of two or more controllers sharing the operation of a large synchronous production process. It should be clear that unnecessary delays or inaccurate information transmitted here will result in lost or damaged production, or even injury to plant personnel. Currently, the unit level is still the domain of proprietary subnetworks, designed by and dedicated to a particular manufacturer of programmable controller. This is changing, but is expected to remain so for some time, as the cost and time effective solutions for this level have yet to evolve as a standard. Figure 12.2 shows a typical factory layout of a plant level network, complemented with a number of shop floor and unit level systems.

12.3 STANDARDS

Standards in the field of communications, like in other fields, allow the development of technological solutions within bounds that encourage a focus of resources and minimize redundancy and waste. The Institute of Electrical and Electronic Engineers (IEEE), has devoted much time and energy to the development of standards for the field of communications. The objective was to develop a standard that would have: (1) a family of media access techniques, (2) one common link control scheme, and (3) one common network management scheme. A diagram illustrating this concept is shown in Figure 12.3. This work on standards began as an outgrowth of Ethernet, pioneered by the Xerox Corporation and others. Ethernet was developed in accordance with

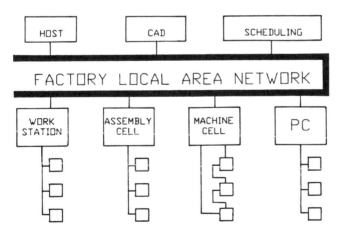

Figure 12.2 Typical factor communication system.

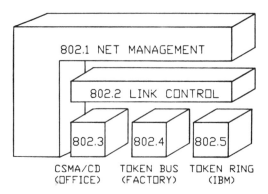

Figure 12.3 IEEE 802 communication structure.

the needs of the modern office. It provided an efficient means for transmittal of large files of information between computers. The IEEE adapted Ethernet characteristics for its current standard for communication, IEEE 802, segmenting it as 802.3 for the office media access technique. This technique is called Carrier Sense Multiple Access/Collision Detection, or CSMA/CD. It is best described as a statistically based scheme, which predicts the availability of the network to all devices equally, but does not guarantee it. As noted, it meets the needs of the office network which is characterized by large files of data sent infrequently.

The factory media access technique under the 802 standard is the Token Bus, and is segmented as 802.4. It provides a deterministic approach to communication, with guaranteed access allowed for all devices equally. It, therefore, provides a real-time synchronous operation with predictable service under heavy loads. It has the disadvantage of a fairly high overhead time to support the token management,and can be unpredictable in nonsteady state conditions. It is this balance of characteristics that makes the Token Bus access scheme most suitable for the plant level networking tasks we examined earlier. It is also being developed for the cell, or shop floor level. Operation of the network allows software to establish a "token" for that device that wishes to initiate a communication session. If no device wishes to initiate, the token continues to move from device to device in a predeter-

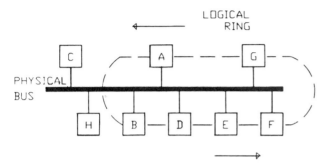

Figure 12.4 Token bus (logical ring) architecture.

mined fashion. This can be a logical ring as shown in Figure 12.4, although the coax arrangement is a bus, or grid. While a device has the token, no other device can initiate a communication. This scheme minimizes collisions on the network. The IEEE 802.4 standard has received a large amount underwriting support from General motors as their Manufacturing Automation Protocol or MAP system.

The Token Ring media access scheme was pioneered by IBM, and is given the segmentation 802.5 in the IEEE standard. It provides for a physical ring arrangement, but is otherwise similar to the Token Bus. The Metro access scheme is segmented as 802.6, and will not be discussed here.

At the time of this writing, most manufacturers of programmable controllers have either adapted their systems to interface on the IEEE 802.4 networking standard, or have specific plans to do so. It is fast becoming the default standard in industrial communications. This is inspite of the fact that the current solution is quite expensive from a per-connect standpoint. As with controllers themselves, costs are expected to fall as the technology spreads.

The rapid emergence of the IEEE 802.4 (MAP) standard is based on accepted or proposed standards, including the International Standards Organization's (ISO), Open System Interconnect Model (OSI), National Bureau of Standards (NBS), and IEEE general standards as noted. The benefit is to provide stability in a multiple vendor environment. The Open System Interconnect

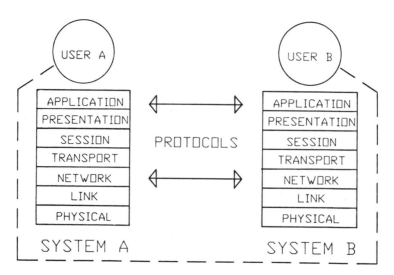

Figure 12.5 Open systems interconnection model.

Model, shown in Figure 12.5, describes in detail the layers of software functionality that must be included in the solution to meet the standard. As the committee continues to refine the description of the standard, manufacturers will modify their equipment to adapt.

12.4 NETWORKS

In this section we will examine some of the component parts of a network, along with the tasks required to properly install and test it. We will consider first the unit level network, then move up to the cell level, and, finally, the plant level.

The unit level network is where most installed programmable controllers find themselves today, if they are connected to any network at all. The communication media here is usually two twisted, shielded pair of conductors, but can be coax in some cases. Data transmission speed is relatively slow, usually under 200,000 bits per second, also expressed as 200 K Baud. Distances covered are modest, generally under 4,000 feet, and connection

scheme flexibility varies from manufacturer to manufacturer.
Some of the more common are shown in Figure 12.6. Structures
include a master-slave, where a single controller is designated as
master, having control over the other units and controlling their
communications. Also common is the peer-to-peer network, also
called the floating master network. It designates a mastership on
a rotating basis, in a manner similar to the token passing scheme
described earlier. Connections to the programmable controller are
made through an interface module, normally residing in the
programmable controller chassis. Figures 12.7 and 12.8 show a
typical interface module separately, and installed in a chassis.
Networks of the unit type normally have limitations in the number of units on the network, usually in the 6 to 10 range. This
occurs because of network throughput problems.

Cell level and plant level will be considered together because
of their many similarities. The chief differences are transmission
speed and cost, although costs at the time of this writing are
similar in some system configurations. This is expected to change

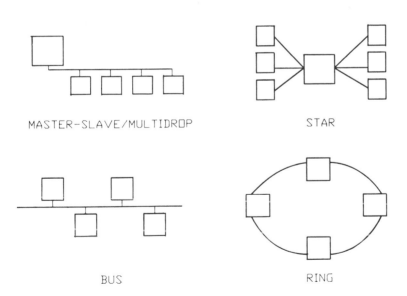

Figure 12.6 Unit level connection schemes.

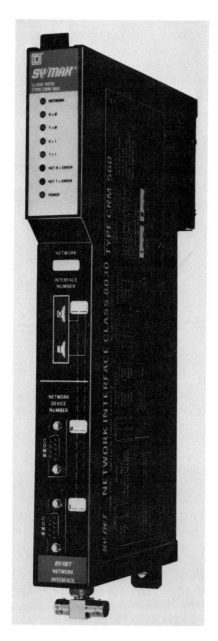

Figure 12.7 Photo of communication interface. (Courtesy of Square D.)

Figure 12.8 Photo of communication interface installed in chassis. (Courtesy of Square D.)

with the gap between the two enlarging with the broadening of the technology.

The transmission media is almost always coaxial cable, although the future may hold enhancements, such as the use of fiber optic media. Transmission speed is 100 K Baud to 1 Megabaud for the cell level, and 5 or 10 Megabuad for the plant level network. This data stream is channeled through a series of cable connections and cable taps to devices called Bus Interface Units (BIUs). These BIUs join the programmable controller interface devices in providing a high speed interface to the network. The BIU takes the frequency based information and converts it to digital data, ready to be decoded. Based on the programmable controller's specific design, the information is then interpreted through protocol conversion to sequences that the controller can interpret and act on. Figure 12.9 shows a typical BIU. The network structure is normally a bus or logical ring. A head-end remodulator is used on broadband systems to reverse the communication direction on the bus, but is not required on baseband systems. And finally, a network manager is an intelligent workstation-like device, which resides on the network and provides overall system configuration management and control. Plant level and cell level networks can range up to 15,000 feet or more in coverage.

While network components play an important role in a well designed communication system, a perhaps more important role is played by the network designer and installer. The design process takes into account the number of devices or clusters that will reside on the network, and their physical positioning in the plant. Choices of cable, taps, and BIUs follow, with many networks having to adapt to existing facility layouts. Once the network hardware is configured and installed, the network certification and debugging process begins. This tests the networks ability to pass information integrally and reliably. The final result is a certified network, ready for programmable controllers and other devices to be connected.

It is important to differentiate between two types of communications software. One is the software that is provided by the communication network provider, and provides basic communication capabilities on the media. Application communication soft-

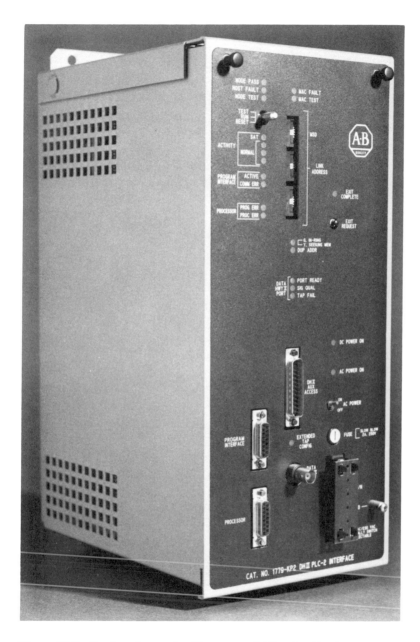

Figure 12.9 Bus interface unit. (Courtesy of Allen Bradley.)

ware designed in a custom fashion specifically for the plant system's needs. For example, it might be the software that set up a communication session to send the current part count from 25 machines to the plant host computer for real-time production monitoring purposes.

13

Future Trends for Programmable Controllers and Their Related Control Systems

The future in industrial control systems is upon us.

In this chapter, the newest methods and trends are examined in light of their likely role in broad control circles. Both hardware and software will be considered, as well as application trends. The picture that begins to take shape will likely be upon us by 1990 or sooner.

13.1 INTRODUCTION

In Chapter 1, we briefly looked at the modest 1970s origins of the programmable controller and its rather narrow application base. Since that humble beginning, these relay replacement devices have benefited from the emergence of the microprocessor, have seen inputs and outputs grow in both variety and span of control distances, and witnessed the use of personal computers and their inherent higher order language capabilities. Throughout this evolution, we have seen costs for equivalent function continue to drop at a rapid pace. Programmable controllers that could sample numerical data, perform local computations, and communicate the results to a distant host computer were not available at any cost in

1970, but are a common purchase item for many manufacturing plants and operations today.

This evolution in programmable controllers has been a small but important part of a larger evolution in today's manufacturing world. This larger evolution, like the smaller one, is driven by a desperate need for improved productivity. Simply stated, today's successful manufacturing business must be a low cost, high quality producer, and must be able to produce low quantities with much the same agility that it does with high volume requirements. This agility comes from having manufacturing apparatus that is flexible in its ability to adapt to a wide variety of product types and lot sizes, that is, it is *programmable*.

Programmable intelligent devices in the factory, office, and engineering areas will combine with efficient communications networks to provide manufacturing businesses that allow all employees to work from the same information sources on a timely basis, minimizing artificial organization barrier problems. By 1990 engineers will design and modify products using the most recent results of customer and market demand trend analysis, communicating the results directly to the manufacturing apparatus after passing through the computer aided scheduling process. Orders from worldwide sales offices will be delivered to manufacturing plants using a wide area communications network. Customer requirements will enter the scheduling process and materials will be ordered and accumulated in a just-in-time fashion, minimizing inventory costs. "Intelligent" production equipment will efficiently produce even small volume custom products, with production flow closely monitored and any inefficiencies triggering alternative production flow paths to materialize. The product is then delivered to an automatic finished goods warehouse, for ultimate shipment to the customer. While this scenario is rare today, some larger firms are already implementing parts of it. And while costs may seem high initially, improvements in productivity are being funded many times by the savings through better management of the inventory process that automation brings. Figure 13.1 shows the likely convergence of design, scheduling and manufacturing.

At the heart of this factory automation lies the once lowly programmable controller. With its powerful and yet general purpose structure, it brings a capability to control manufacturing

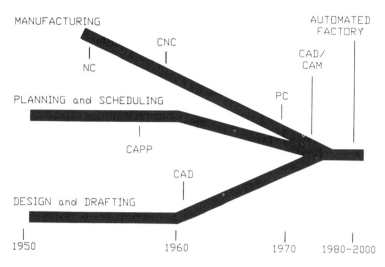

Figure 13.1 Diagram of converging technologies over time.

processes reliably and efficiently, combined with the ability to collect and communicate information rapidly. It is this device that we will examine as it might look in 1990.

13.2 HARDWARE

13.2.1 Central Processing Unit

The hardware used up until this time has changed dramatically since the first devices in 1970. The invention and application of the microprocessor insured this. Eight-bit microprocessors were the latest in the middle to late 1970s, with the 16-bit following in the 1980s. Bit slice architectures were common in the first half of the 1980s decade, but by 1990 things will likely change again.

The programmable controller of the year 1990 may not even be called a programmable controller. A more likely term might be the unit controller, signifying where the controller lies in the control hierarchy rather than its traditional characteristics. Indeed, 1990's controller will almost certainly take advantage of the winner of today's "standards wars," between popular competing

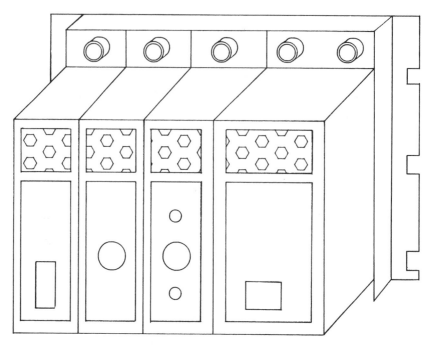

Figure 13.2 Example of emerging standard control hardware.

bus standards. IBM's PC Bus, Intel's Multibus II, Pro-Log's STD
Bus, and Motorola's VMeBus are the leading contenders at this
point. A possible example of how this might look is shown in
Figure 13.2. These standard will allow use of more powerful 32-
bit microprocessors and multiple processing paths. Multiple pro-
cessors will still be used, but tasks will be more clearly defined and
operating systems will support multiple application execution in a
standard fashion. The other major attribute that this new unit
controller will have is the ability to share memory with other controllers. This will allow easier integration of a global manufacturing data base at the cell level. This open structure will encourage
third party manufacture of the more common memory and communication subsystems, while making the traditional programmable controller manufacturer into more of a specialized system
integrator whose primary value added is software development and

integration, along with industry application expertise, installation skills, and service excellence for these new unit controllers and associated equipment. Utilizing this multiple path system will be multiple scans operating in parallel. This will ensure the efficiency of the unit controller for a variety of tasks, well beyond the current programmable controller functional envelope.

As a part of this evolution, cell and area controllers may look suspiciously like unit controllers. This standardization, while fierce, will allow users to train their maintenance personnel on common equipment, regardless of the manufacturer. Also aiding this process will be standard high performance diagnostics capabilities in the new unit controller. This will allow the user to perform common maintenance on most of his equipment, enhancing the prospects of higher productivity through less downtime. These diagnostic capabilities will be designed to mesh with diagnostic capabilities in the I/O systems. Data on the specific nature of the fault will be collected, analyzed, and reported automatically in a standard way that is common for all controllers. Trend data will be compared to that of expected failure frequency, the alarming conclusions will be reported. This analysis will make use of expert-type systems to accomplish this.

13.2.2 Input/Output Systems

In 1990, the I/O systems associated with programmable controllers will be very powerful compared to today's standards. The ability to interact with a variety of sensors and actuators in a highly intelligent fashion will be common. Diagnostics will be readily available, and will be used in a fashion that is commonly understood by system designers and maintenance personnel alike. I/O will be further distributed in nature than it is today. In fact, if today's trends continue, much of the I/O of tomorrow will actually be integrated into the sensors and actuators themselves, thereby removing the interim step of I/O blocks or chassis. I/O blocks that are used may contain rudimentary logic allowing local control to be executed. The CPU in this instance acts as supervisor, overriding the local block occasionally in nonroutine circumstances. With continued progress in standards, the I/O of 1990 might even be plug compatible with a variety of unit controller CPU brands, allowing very aggressive growth in this seg-

ment. Just like some of the I/O systems of today, tomorrow's
I/O will be programmable on a per-point basis. The ability to
configure a point as input or output along with adjustable current
thresholds and circuit protection will be common. Diagnostics
will be provided to allow detection and automatic reporting of
faults in a standard manner to the CPU. Short and open circuit
conditions will be pinpointed quickly, and recommended action
will be highlighted for the maintenance dispatch.

13.2.3 Programming Devices

Following the current trend, programming devices of tomorrow
will be spin-offs and adaptations of today's personal computer
revolution. Industrialized versions of these powerful devices will
flourish on the factory floor, and will be used not only to program
the programmable (unit controllers), but will program all of the
intelligent devices in the factory, including robot controllers, computer numerical controllers, variable speed drive systems, and
vision systems controllers. These factory floor programmers will
exist in both portable forms and forms that are dedicated as a
node on the factory's local area network. Operating system software will exist as standards similar to that of today, but application software will be widely differentiated, again similar to that of
today's office software. Favorites will emerge as default standards. One likely occurrence for programmers is the ability to
create a program on a programmer, and then simulate the execution of that program on the programmer as well. This added step
will allow the efficient creation, testing, and debugging of programs in parallel, or well before the unit controller hardware is
purchased or configured.

13.2.4 Operator Interfaces

The last five years have seen rapid growth in the development and
use of intelligent operator interfaces. It is likely that the next five
years will see even faster growth. The interaction that an operator
or system designer will have with programmable controllers in
1990 will probably come in forms that are logical and natural.
Included will be color image-based interactions combined with
computer-generated voice outputs. Inputs will consist of touch

or other point systems, along with voice input on some applications. As a natural by-product of hardware and software standards developments, message and other information formats will be more common from system to system, allowing many different users to be instantly familiar with the system.

13.3 SOFTWARE

Just as software is playing a major role today, it will continue to play a significant role in the programmable controllers of 1990. Of importance to note are trends in both operating system software and application software.

13.3.1 Operating Systems

In most of today's programmable controllers, the operating systems are proprietary in nature. That is, they are designed by the controller manufacturer to operate in an optimum but nonstandard way. Just as standards have evolved on current generations of personal computers, it is likely that the controller of 1990 will have the ability to choose among a small number of standard operating systems. In fact, even today, a number of good operating systems exist and are competing to become standards. Some of these are of the multitasking variety, allowing several applications to run concurrently. Unix is an example of a possible real-time operating system candidate that is currently enjoying a wave of popularity.

13.3.2 Languages

Programmable controller languages will define the true meaning of the phrase user friendly by 1990. Software developers, both in the programmable controller manufacturer's firm, and third party, will branch off from standard operating systems to develop a dazzling array of application software. The use of highly developed forms of operator interface will be commonplace, and application programs will be easily configured for specific industrial applications. The use of graphical techniques for the purpose of communicating to system designers and operators will be extremely powerful. They will circumvent any problems of interpretation by a wide variety of users. A language structure is likely to emerge by 1990 or before that will include the evolution of ladder logic, function block, and Boolean. Combinations or conversions of this language will allow system integrators, equipment manufacturers, and users to benefit from a multilevel approach, similar to that shown in Figure 13.3.

13.4 COMMUNICATIONS

The communications genesis of the controller of 1990 are clearly evident today. Nonproprietary local area networks and high performance subnets, as well as proprietary subnets are in development stages today, with costs and functionality in all phases of evolution. The next few years promise continued progress toward the ultimate goal of a few good standards that many vendors' equipment can adapt to easily. This will include both a settling of hardware and software standards, with the latter perhaps the more formidable challenge.

We saw earlier references to the Manufacturing Automation Protocol, or MAP standard for industrial communication. MAP is likely to continued a rapid evolution into a widely accepted and used standard for factory "backbone" communications. That is, it will provide basic services for a large number of unrelated intelligent devices to communicate with. While the trend is not as clear, it follows logically that existing standards for office communications will merge toward a common interchange capability with the successful industrial communication standard(s). A likely contender for at least part of the task is the Technical and Office Protocol, or TOP standard. At the time of this writing, it has been successfully modeled with the MAP system. This hybrid communication system offers the benefits of a common method to move information created in an office or engineering design system directly to the factory floor and on to the production machinery. Future enhancements will permit managers of the manufacturing business to share three-dimensional images that convey information among many widespread users. This same information will be easily converted to that needed for factory floor use— in the year 1990.

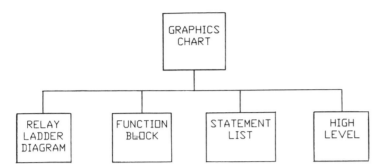

Figure 13.3   Diagram of a multiple language evolution structure.

Another concept to consider in the field of industrial communications is that of using unbounded media, or airwaves as a communication medium. Possible enhancements in radio, microwave, or infrared technologies may allow radical changes in the way industrial communications systems are designed and used. They may also bring with them a new level of flexibility and cost effectiveness.

13.5   APPLICATIONS

Speculation in this chapter leads one to believe that the controller design of 1990 will be a blend of many of today's best technologies. The positive attributes of programmable controllers, general purpose computers, and dedicated microprocessor-based controllers will contribute to a powerful new form of unit controller.

With this in mind, it is logical that applications for these new controllers will be a blend of many of today's as well. Programmable controllers are today being used for controlling all types of production machinery, from the very smallest stand-alone machine to the largest batch process. Yes, every conceivable product, from thimbles and aircraft engines to medicines and plastic film can be made with the help of programmable controllers. Considering the breadth of the current application picture, it is hard to imagine its scope being any broader. But with the controller (and supporting technologies) of 1990, it is likely to be broader indeed. Many machines and processes that were previously considered poor candidates for automation because of cost or complexity will join the fold as functionality per dollar of cost continues to rise. In addition, many light industrial and commercial applications will become candidates for programmable controller automation.

The application scene of 1990 will be one of very specific and sharp focus on systems and industry specific solutions. The first signs of this are appearing on the horizon just now. The ease of use of tomorrow's controller will be facilitated by continuing enhancements in the area of programming languages. Languages will continue to evolve using higher order graphics programming techniques, which will be usable by a wider and wider category of plant personnel. This friendliness will pave the way for application program developments that are customized for very specific application and industry segments. For example, it should be common by 1990 to be able to purchase a hardware and software combination that is easily configured as a multiple axes motion control system, programmed in the designers choice of common languages, i.e., English, German, Japanese, etc. In addition, the designer will be able to choose the engineering units most appropriate, English or metric. Once a system is designed, integrated, and shipped to the installation site, users will be able to easily make the system operational through multiple language operator prompts. Maintenance will be easier as well since standard hardware will be supplemented with common diagnostic routines, pinpointing problems and recommending repair action.

13.6 SUMMARY

We have traced the heritage of the programmable controller from its humble beginning as an awkward relay replacement vehicle. From a slow start, it took on ever more functions and capabilities as it shrunk in size and cost. After mastering simple sequencing it tackled motion control, then process control, data management, and communications. And while it is likely that the controller will change as time marches forward, it is not clear at all that its achievements have plateaued or progress has been slowed. Indeed if any lesson can be extracted from the first 15 years of the programmable controller's existence, it is that creative users will continue pushing back the envelope of application to new and wider vistas.

Bibliography

1. General Electric Company, *Genius Input Output System*, GED7003, Charlottesville, Virginia (1985).

2. Gschwind, Hans W. *Design of Digital Computers*, Springer-Verlag, Wein, New York (1971).

3. Jones, Clarence T. and Bryan, Luis A. *Programmable Controllers: Concepts and Applications*, International Programmable Controls Inc., Atlanta, Georgia (1983).

4. Curtin, Keith. Simplifying PC Troubleshooting and Maintenance Jobs, *Plant Engineering*, (January 24, 1985) pp. 84-86.

5. Lloyd, Mike. Graphical Function Chart Programming for Programmable Controllers, *Control Engineering*, (October, 1985).

6. General Electric Company, *Series Six Programmable Controllers Product Summary*, GET6748, Charlottesville, Virginia (1984).

7. General Electric Company, *Series Six Programmable Controller Installation and Maintenance Manual*, GEK25361A, Charlottesville, Virginia (1982).

8. General Electric Company, *Series Six Programmable Control-*

ler Axis Positioning Manual, GEK25368, Charlottesville, Virginia, (1984).

9. General Electric Company, *Series Six Programmable Controller ASCII/Basic Module Manual*, GEK25398, Charlottesville, Virginia, (1984).

10. General Electric Company, *Series Six Programmable Controller Data Communication Manual*, GEK25364, Charlottesville, Virginia (1984).

11. General Electric Company, *Series Six Programmable Controller Programming Manual*, GEK25362, Charlottesville, Virginia (1982).

12. General Electric Company, *Series Six Programmable Controller Operator Interface Terminal Manual*, GEK90817, Charlottesville, Virginia (1984).

13. General Electric Company, *Series Six Programmable Controller Proloop System*, GET6844, Charlottesville, Virginia (1984).

14. General Electric Company, *Series Six Programmable Controller Redundant Processor Manual*, GEK25366A, Charlottesville, Virginia (1984).

15. General Electric Company, *Series Six Programmable Controller Application Guide*, GEK25365A, Charlottesville, Virginia (1984).

16. Square D Company, *Sy/Max Class 8030 Type RIM-131 High Speed Counter Module Manual*, Bulletin 30598-210-01, Milwaukee, Wisconsin (1984).

17. Square D Company, *Sy/Max Class 8010 PID Closed Loop Control Using the Sy/Max Model 300 Programmable Controller*, Bulletin 30598-301-02, Milwaukee, Wisconsin (1984).

18. Square D Company, *Sy/Max Class 8030 Type CRM-510 Sy/Net Network Interface Module*, Bulletin 30598-257-01, Milwaukee, Wisconsin (1984).

19. Square D Company, *Sy/Max Class 8010 Type SPR-300,310 Deluxe Programmer*, Bulletin 30598-167-01, Milwaukee, Wisconsin (1984).

20. Square D Company, *Sy/Max Programmable Controller Planning and Installation Guide*, Bulletin 30598-175-01, Milwaukee, Wisconsin (1985).

21. Square D Company, *Sy/Max Color Operator Interface*, Bulletin SM679, Milwaukee, Wisconsin (1985).

22. Wiatroski, Claude A. and House, Charles H. *Logic Circuits and Microcomputer Systems*, McGraw-Hill Book Company, New York (1980).

23. General Electric Company, *Workmaster Programmable Controller Information Center Guide to Operation*, GEK25373, Charlottesville, Virginia (1984).

24. General Electric Company, *Workmaster Programmable Controller Information Center Logicmaster 6 Programmer Documenter Manual*, GEK25379, Charlottesville, Virginia (1984).

Appendixes

A
Glossary of Terms

Actuator: Category of field device connected to I/O.

Address: Number used to define a specific memory location in the programmable controller.

Alphanumeric: Any combination of alphabet, numerals, or other characters used for representing information.

Analog signal: A signal having a smoothly and continuously varying quality as contrasted to discrete changes.

AND: A Boolean operator that causes logic 1 to occur if, and only if, all inputs are 1, otherwise, logic 0 occurs.

ANSI: American National Standards Institute.

Application Program: Coded language, normally created by user or system integrator, that causes specific control actions to occur in a predictable fashion. This program is stored in application memory.

ASCII: American Standard Code for Information Interchange.

Assembly language: A lower level, symbolic language that directly converts to machine language.

Asynchronous communication: A method that uses start and stop units to define the beginning and end of characters, allowing uneven time intervals between communications.

B

Baseband: A communication method that allows individual signals, either modulated or unmodulated to occur.

Baud: The number of binary bits per second transmitted, in a serial communication.

Binary Coded Decimal (BCD):   A number system where each decimal digit from 0 to 9 is represented by four binary bits.

Binary number: A number system that uses 0 and 1 in combinations to describe information for use by digital machines.

Bit: A binary digit; also the smallest element of binary information.

Boolean: A system of logic that uses operators such as AND, OR, NAND, NOR, and Exclusive OR, in arrangements that provide true or false outputs.

Broadband: Communications of more than one stream of information simultaneously through frequency multiplexing.

Bus: An arrangement of conductors used for data communication and/or control.

Byte: A group of eight adjacent bits.

C

Cathode Ray Tube (CRT):   A component used in programming devices and operator interface systems.

Central Processing Unit (CPU):   In a programmable controller, it is the primary intelligence used for arithmetic and timing operations.

Checksum: A technique used to check data transmission integrity that places a character at the end of a data block.

Complimentary Metal Oxide Semiconductor (CMOS):   Integrated circuits that have high noise immunity and consume little power.

Coaxial Cable: A physical communication line that uses a single conductor with concentric insulator and braided shield.

Contact: A conductor in an electromechanical device. It can be represented by symbols for use in a programmable controller as normally open or normally closed.

Contention system: A communication plan that allows common status to all devices, defining contention to access the bus.

Control logic: See Application program.

Core Memory: Memory type that uses toroidal coils electrically excited to store binary information.

Cyclic Redundancy Check (CRC):   A data integrity checking scheme that uses a binary number to manipulate a block of data, checking division, and remainder.

D

Data highway: A system which allows data from separate intelligent devices to be passed one to the other in a prescribed manner.

Debounce: The design procedure that removes non-steady state signals from the logic input in electromechanical switches.

Debug: Test of a completed system for the purpose of finding and eliminating undesirable characteristics and performances.

Diagnostics: Features in a control system that provide automatically the ability to detect and report certain fault conditions.

Distributed control: A design philosophy in control systems that provides for individual intelligent devices to be configured into a single large system using various data communications schemes.

Documentation: The descriptions, diagrams, and application software necessary to install, use, and maintain the system in a programmable controller based system.

E

EAROM: Electrically Alterable Read Only Memory.

EEPROM: Electrically Eraseable Programmable Read Only Memory.

EIA: Electronics Industries Association.

EPROM: Eraseable Programmable Read Only Memory.

Executive Memory: The system memory, normally in a read only form, that provides the basic intelligence for the system, establishing operating characteristics.

F

File: A block of information segmented and treated as a unit.

Flag: A system software practice that establishes bits of binary information to establish the occurrence of an event.

Flow chart: A software design technique that allows the programmer to establish various sequences that are to become programmed events.

Full duplex: A communication where information flows in both directions simultaneously.

G

Gate: A logical device whose output is activated in a manner described with its logical function.

Global I/O: Inputs and Outputs whose bit status are available to other programmable controllers through shared data tables.

Gray Code: A cyclic numbering code used primarily with position sensing systems.

H

Half duplex: A communication where information flows in one direction at a time.

Hard copy: Usually the printed form of computer generated information.

Hardware: Physical programmable controller items, along with peripheral devices.

Header: Part of the software used to format and permit communication of information files.

Hexadecimal Number System: Numbering system that uses as its base the number 16, articulating representations with the numbers 0 through 9, and the letters A through F.

Higher Level Language: A programming language that uses a powerful instruction set that breaks down to executable algorithms in machine execution.

Host computer: A device that provides broad services to the other devices present, including data collection, analysis, routing, and storage in a manufacturing system.

I

Information rate: A unit used to define communication speed and efficiency.

IEEE: Institute of Electrical and Electronic Engineers.

Instruction set: A particular group of coded instructions that provide a specific, intelligent, method to deal with the machine.

Intelligent terminal: A subsystem used to interact with the programmable controller that provides local intelligence. This may be in the form of CRT screens of graphics and/or dynamic data.

Internal output: An output in the programmable controller I/O structure that is used exclusively for software or "internal" logical purposes.

I/O: Input/Output.

I/O Address: A connotation used in the I/O structure to assign a specific number to an I/O point or channel.

I/O Module: A fabricated and assembled group of circuit board(s) and components designed to reside in the programmable controller I/O chassis.

I/O update: The process of refreshing the input and output structure of the programmable controller to bring in new all input status and send out all output status. The interval required for this is called the I/O update time or I/O scan time.

Isolated I/O: A special class of discrete I/O that uses an individual common for each input or output point. This design allows a variety of field device types to be used and provides for a convenient way to disconnect power to the I/O locally without generating a fault.

L

Ladder diagram: A programmable controller language that uses contacts and coils to define a control sequence.

LAN: *See* Local Area Network.

Latch: An instruction used in ladder diagram programming to represent an element that retains its state during controlled toggle and power outage.

LRC: Longitudinal Redundancy Check.

LSB: Least Significant Bit.

Local Area Network: A system of hardware and software designed to allow a group of intelligent devices to communicate within a fairly close proximity, i.e., 2 miles.

M

Machine language: A control program reduced to binary form.

Memory: The part of the programmable controller that contains the control program and other temporary information.

Mnemonic codes: Symbols that are designated to represent a specific set of instructions for use in a control program.

MSB: Most Significant Bit.

Multiplex: The act of utilizing one channel of I/O of the controller to share input and output information.

N

NAND: The logic gate that result in 1 unless both inputs are 0.

NEMA: National Electrical Manufacturers Association.

Network: A system that is connected for communication purposes.

Node: A point on the network that allows access.

Noise: Spurious interference in a programmable controller or network.

Nonvolatile memory: A type of memory in a controller that does not require power to retain its contents.

NOR: The logic gate that results in 0 unless both inputs are 0.

NOT: The logic gate that results in the complement of the input.

O

Octal: Number system based on the number 8, utilizing numbers 0 through 7.

Operating system: The fundamental software for a system that defines how it will store and transmit information.

Optical isolation: A technique used in I/O module design that provides logic separation from field levels.

OR: The logic gate that results in 1 unless both inputs are 0.

P

Parallel: A system whereby groups of bits are communicated simultaneously.

Parity: A communications checking device convention that uses odd or even numbers of bits to provide data integrity through comparisons.

Peer-to-peer; A communication system that allows all devices to initiate as well as to receive messages.

PID: Proportional Integral Derivative.

PLC: Programmable Logic Controller.

Programmable controller: A system of hardware and software whose basic operation is to solve logic repetitively, is programmable in ladder logic, and is suitable for installation in industrial environments.

PROM: Programmable Real Only Memory.

Protocol: Communication conventions defining how certain communication occur.

R

RAM: Random Access Memory.

Redundancy: The programmable controller system configured to increase system availability through multiple modules or systems.

Register: A temporary storage location in programmable controller memory.

Relay: An electromechanical device utilizing an adjacent coil and contact.

ROM: Real Only Memory.

Rung: Convention used in referring to ladder logic programs.

S

Scan time: The amount of time needed to complete one programmable controller cycle of memory and I/O update.
Sensor: A category of field device connected to I/O.
Serial communication: A method that provides for single bits
of information to be transmitted.

T

Throughput: The speed in which information flows through a
system.
Thumbwheel: A component used to allow operators to input to
the programmable controller specific numeric data.
Topology: A method used in arranging devices in a network.
Truth table: A listing showing outputs for all possible inputs.

U

User memory: That segment of memory that is available for
application programs.

V

Volatile Memory: A type of memory that loses its contents upon
loss of power.

W

Watchdog timer: An internal checking mechanism that ensures
that the programmable controller system is operating cor-
rectly.
Word: A number of bits combined to form a basic memory
element, referring to controller memory.

B
Abbreviations and Acronyms

AC	Alternating Current
ANSI	American National Standards Institute
ASCII	American Standard Code for Information Exchange
BCD	Binary Coded Decimal
BIU	Bus Interface Unit
CMOS	Complimentary Metal Oxide Semiconductor
CNC	Computer Numerical Control
CPU	Central Processing Unit
CRC	Cyclic Redundancy Check
CRT	Cathode Ray Tube
CSMA/CD	Carrier Sense Multiple Access/Collision Detect
DC	Direct Current
EAROM	Electrically Alterable Read Only Memory
EEPROM	Electrically Eraseable Programmable Read Only Memory
EIA	Electronic Industries Association
EMI	Electromagnetic Interference
EPROM	Eraseable Programmable Read Only Memory
FIFO	First in First Out
HDLC	High (level) Data Link Control
IEC	International Electrical Commission
IEEE	Institute of Electrical and Electronic Engineers
I/O	Input/Output

IOR	Inclusive OR
ISO	International Standards Organization
LAN	Local Area Network
LCD	Liquid Crystal Display
LED	Light Emitting Diode
LIFO	Last In First Out
LRC	Longitudinal Redundancy Check
LSB	Least Significant Bit
mA	milli-Amp
MAP	Manufacturers Automation Protocol
MCR	Master Control Relay
MODEM	Modulater/Demodulater
MSB	Most Significant Bit
msec	millisecond
MTBF	Mean Time Between Failures
MTTR	Mean Time To Repair
NC	Numerical Control
NEMA	National Electrical Manufacturers Association
NOVRAM	Non-Volatile Random Access Memory
OEM	Original Equipment Manufacturer
OSI	Open System Interconnect
PC	Programmable Controller (or Personal Computer)
PID	Proportional Integral Derivative
PLC	Programmable Logic Controller
RAM	Random Access Memory
RF	Radio Frequency
ROM	Read Only Memory
RTD	Resistive Thermal Device
R/W	Read/Write
TOP	Technical and Office Protocol
TTL	Transistor Transistor Logic
UPS	Uninterruptible Power Supplies
XMIT	Transmit
XOR	Exclusive OR

C
ASCII Cross References

Decimal	Hex	ASCII character	Controller equivalent
001	01	NUL	A
002	02	STX	B
003	03	ETX	C
004	04	EOT	D
005	05	ENQ	E
006	06	ACK	F
007	07	BEL	G
008	08	BS	H
009	09	HT	I
010	OA	LF	J
011	OB	VT	K
012	OC	FF	L
013	OD	CR	M
014	OE	SO	N
015	OF	SI	O
016	10	DLE	P

ASCII CROSS REFERENCES (continued)

Decimal	Hex	ASCII character	Controller equivalent
017	11	DC1	Q
018	12	DC2	R
019	13	DC3	S
020	14	DC4	T
021	15	NAK	U
022	16	SYN	V
023	17	ETB	W
024	18	CAN	X
025	19	EM	Y
026	1A	SUB	Z
027	1B	ESC	[
028	1C	FS	\
029	1D	GS	]
030	1E	RS	
031	1F	US	
032	20	SP	
033	21	!	
034	22	''	
035	24	#	
036	25	$	
037	26	%	
038	27	&	
039	28	'	
040	29	(	
041	29	)	
042	2A	*	
043	2B	+	
044	2C	'	
045	2D	—	
046	2E	-	
047	2F	/	

ASCII CROSS REFERENCES (continued)

Decimal	Hex	ASCII character	Controller equivalent
048	30	0	
049	31	1	
050	32	2	
051	33	3	
052	34	4	
053	35	5	
054	36	6	
055	37	7	
056	38	8	
057	39	9	
058	3A	:	
059	3B	;	
060	3C	<	
061	3D	=	
062	3E	>	
063	3F	?	
064	40	@	
065	41	A	
066	42	B	
067	43	C	
068	44	D	
069	45	E	
070	46	F	
071	47	G	
072	48	H	
073	49	I	
074	4A	J	
075	4B	K	
076	4C	L	
077	4D	M	
078	4E	N	

ASCII CROSS REFERENCES (continued)

Decimal	Hex	ASCII character	Controller equivalent
079	4F	O	
080	50	P	
081	51	Q	
082	52	R	
083	53	S	
084	54	T	
085	55	U	
086	56	V	
087	57	W	
088	58	X	
089	59	Y	
090	5A	Z	
091	5B	[	
092	5C	\	
093	5D	]	
094	5E		
095	5F	—	
096	60	Accent Grave	
097	61	a	
098	62	b	
099	63	c	
100	64	d	
101	65	e	
102	66	f	
103	67	g	
104	68	h	
105	69	i	
106	6A	j	
107	6B	k	
108	6C	l	
109	6D	m	

ASCII CROSS REFERENCES (continued)

Decimal	Hex	ASCII character	Controller equivalent
110	6E	n	
111	6F	o	
112	70	p	
113	71	q	
114	72	r	
115	73	s	
116	74	t	
117	75	u	
118	76	v	
119	77	w	
120	78	x	
121	79	y	
122	7a	z	
123	7B		
124	7C	:	
125	7D		
126	7E	~	
127[a]	7F	Delete	

[a]ASCII Table continues to decimal 255, and includes codes for various symbols and graphic characters.

D

Number Tables

Powers of Two (Binary)

n	2^n
0	1
1	2
2	4
3	8
4	16
5	32
6	64
7	128
8	256
9	512
10	1,024
11	2,048
12	4,096
13	8,192
14	16,384
15	32,768
16	65,536

Powers of Eight (Octal)

n	8^n
0	1
1	8
2	64
3	512
4	4,096
5	32,768
6	262,144
7	2,097,152
8	216,777,216
9	134,217,728
10	1,073,741,824
11	8,589,934,592
12	68,719,476,736
13	549,755,813,888
14	4,398,046,511,104
15	35,184,372,088,832
16	281,474,976,710,656

Powers of Sixteen (Hexadecimal)

n	16^n
0	1
1	16
2	256
3	4096
4	65,536
5	1,048,576
6	16,777,216
7	268,435,456
8	4,294,967,296
9	68,719,476,736
10	1,099,511,627,776
11	17,592,186,044,416
12	281,474,976,710,656
13	4,503,599,627,370,496
14	72,057,594,037,927,936
15	1,152,921,504,606,846,976
16	18,446,744,073,709,588,616

E

Electrical Diagram Symbols

SWITCHES	PUSH BUTTONS	MOTORS AND INDICATING LIGHTS	OTHER
LIMIT N. O.	SINGLE CIRCUIT N. O.	INDICATING LIGHT	FUSE
LIMIT N. C.	SINGLE CIRCUIT N. C.	MOTOR	OVERLOAD RELAY

OPERATING COILS		RELAY AND AUXILIARY CONTACTS	CONTACTOR AND STARTER POWER CONTACTS	
MAIN	UNLATCH	NORMALLY OPEN	NORMALLY OPEN	TIME DELAY N.C.
START	REVERSE	NORMALLY CLOSED	NORMALLY CLOSED	TIME DELAY N.O.
FORWARD	NORMAL			
CONTROL	TIMING			

F
Manufacturers and Their Addresses

Asea Industrial Systems
16250 West Glendale Dr.
New Berlin, WI 53151

Adatek, Inc.
1223 Michigan
Sandpoint, ID 83864

Allen Bradley
(Div Rockwell Intl.)
747 Alpha Drive
Highland Heights, OH 44143

Automation Systems
208 No. 12th Ave.
Eldridge, IA 52748

Bailey Controls Co.
29801 Euclid Ave.
Wickliffe, OH 44092

Cincinnati Milacron
Mason Rd. & Rte. 48
Lebanon, OH 45036

Devilbiss Corp.
9776 Mt. Gilead Rd.
Fredricktown, OH 43019

Eagle Signal Controls
8004 Cameron Rd.
Austin, TX 78753

Eaton Corp.—Cutler Hammer
4201 North 27th St.
Milwaukee, WI 53216

Eaton Leonard Corp.
6305 El Camino Real
Carlsbad, CA 92008

Foxboro Co.
Foxboro, MA 02035

Furnas Electric
1000 McKee St.
Batavia, IL 60510

GEC Automation Projects
2870 Avondale Mill Rd.
Macon, GA 31206

General Electric Co.
PO Box 8106
Charlottesville, VA 22906

General Numeric
390 Kent Ave.
Elk Grove Village, IL 60007

Giddings & Lewis
Electr. Div.
666 South Military Rd.
Fond du Lac, WI 54935-7258

Gould Inc.
Programmable Control Div.
PO Box 3083
Andover, MA 01810

Guardian/Hitachi
1550 W. Carroll Ave.
Chicago, IL 60607

Honeywell
IPC Div.
435 West Philadelphia St.
York, PA 17404

Klockner--Moeller
Natick, MA

Maxitron Corp.
Salem, NH

McGill Mfg. Co.
Elec. Div.
1002 N. Campbell St.
Valparaiso, IN 46383

Mitsubishi Electric
799 N. Bierman Circle
Mt. Prospect, IL 60056-2186

Modular Computer Systems
Inc.
1650 W. McNabb Rd.
Fort Lauderdale, FL 33310

Omron Electric
Control Div
One East Commerce Drive
Schaumburg, IL 60195

Reliance Electric
Centrl. Sys. Div
4900 Lewis Rd.
Stone Mountain, GA 30083

Siemens—Allis Automation
Inc.
10 Technology Drive
Peabody, MA 01960

Square D Co.
4041 N. Richards St.
Milwaukee, WI 53201

Struthers-Dunn Systems Div.
4140 Utica Ridge Rd.
Bettendorf, IA 52722

Telemecanique
901 Baltimore Blvd.
Westminster, MD 21157

Texas Instruments
Indl. Control Dept
PO Drawer 1255
Johnson City, IN 37605-1255

Toshiba
13131 West Little York Rd.
Houston, TX 77041

Triconex
16800 Aston St.
Irvine, CA 92714

Westinghouse Elec. Numa—
Logic
1512 Avis Drive
Madison Heights, MI 48071

G
Standards and Committees

American National Standards Committee (ANSI)
1430 Broadway
New York, New York 10018

Electronic Industries Association (EIA)
2001 I Street NW
Washington, D.C. 20006

Institute of Electrical and Electronic Engineers (IEEE)
345 East 47th St.
New York, New York 10017

Instrument Society of America (ISA)
67 Alexander Drive
Research Triangle Park, North Carolina 27709

International Standards Organization (ISO)
(Division of ANSI)
1430 Broadway
New York, New York 10018

National Electrical Manufacturers Association (NEMA)
2101 L. Street NW
Washington, D.C. 20037

Society of Manufacturing Engineers (SME)
P. O. Box 930
One SME Drive
Dearborn, Michigan 48121

H
Boolean Symbols and Formulas

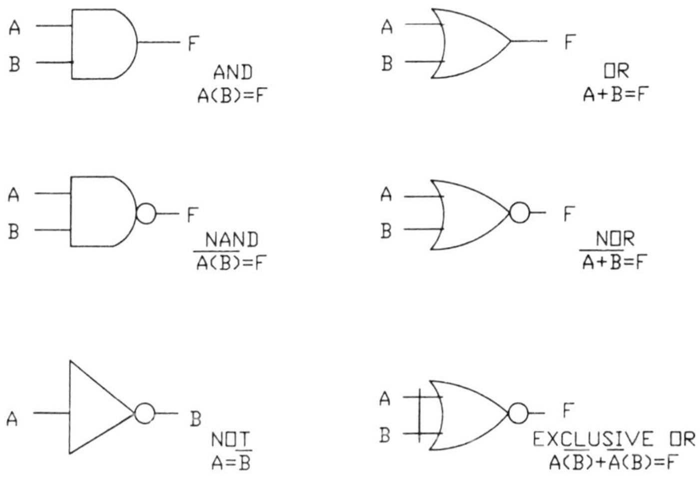

A
B
F
AND
A(B)=F

A
B
F
OR
A+B=F

A
B
F
NAND
$\overline{A(B)}$=F

A
B
F
NOR
$\overline{A+B}$=F

A
B
NOT
A=$\overline{B}$

A
B
F
EXCLUSIVE OR
$\overline{A}$(B)+A($\overline{B}$)=F

Index

DAT